T0269064

Deuterium

Deuterium

Discovery and Applications in Organic Chemistry

Jaemoon Yang

AMSTERDAM • BOSTON • HEIDELBERG • LONDON • NEW YORK • OXFORD
PARIS • SAN DIEGO • SAN FRANCISCO • SINGAPORE • SYDNEY • TOKYO

Elsevier
Radarweg 29, PO Box 211, 1000 AE Amsterdam, Netherlands
The Boulevard, Langford Lane, Kidlington, Oxford OX5 1GB, UK
50 Hampshire Street, 5th Floor, Cambridge, MA 02139, USA

British Library Cataloguing-in-Publication Data
A catalogue record for this book is available from the British Library

Library of Congress Cataloging-in-Publication Data
A catalog record for this book is available from the Library of Congress

ISBN: 978-0-12-811040-9

For Information on all Elsevier publications
visit our website at http://www.elsevier.com/

Publisher: John Fedor
Acquisition Editor: Katey Birtcher
Editorial Project Manager: Jill Cetel
Production Project Manager: Anitha Sivaraj
Designer: MPS

Typeset by MPS Limited, Chennai, India

DEDICATION

To Urey and those who follow.

CONTENTS

ACKNOWLEDGMENTS

In the preparation of this book, I have received a great deal of assistance from many people. Without them this book would not have been possible.

I am pleased to acknowledge help from Professor Michael Krische of The University of Texas at Austin and Professor Takahiko Akiyama of Gakushuin University in Japan, who provided me with additional information on their research.

I would like to express my sincere appreciation to the following individuals at Cambridge Isotope Laboratories, Inc. (CIL): Dr. Joel Bradley instilled a great value of deuterium in me. Dr. William Wood encouraged me to explore a wonderful world of deuterium. Dr. Richard Titmas read the entire manuscript and made valuable suggestions. Mrs. Diane Gallerani obtained a number of rare articles in a timely manner. My colleagues at CIL, Drs. Sun-Shine Yuan, Susan Henke, Steven Torkelson, and Salim Barkallah, are gratefully acknowledged for their proofreading efforts and helpful comments.

I wish to thank the Elsevier publishing team for making the manuscript become a book. Katey Birtcher, senior acquisitions editor of chemistry, kindly accepted the book proposal and arranged a very smooth review process. Jill Cetel, senior editorial project manager, was instrumental ensuring that the book was ready to print.

Finally, I want to thank my wife, Wenjing Xu, PhD, for her support and love.

Jaemoon Yang, Ph.D.

Cambridge Isotope Laboratories, Inc.
Andover, MA

April 2016

INTRODUCTION

HYDROGEN IS UBIQUITOUS

It is everywhere around us. The water we drink every day is made up of hydrogen and oxygen, the gasoline we pump at the gas station contains hydrogen and carbon, the sugar we use is made of hydrogen, carbon, and oxygen. DNA is another fine example: it has hydrogen as well as other atoms such as carbon, nitrogen, oxygen, and phosphorus. Recently, hydrogen-fueled vehicles are gaining attention as a zero-emission alternative. Being the lightest of all the elements in the Periodic Table, hydrogen is one of the most common atoms that make up the world.[1]

Since its discovery in 1766 by Henry Cavendish, hydrogen had been considered a pure element for more than 160 years. It turned out that hydrogen is not pure! It is very close to being pure, but it is not exactly 100%. The chemical purity of hydrogen is 99.985%. This is because there are two forms or isotopes of hydrogen: the protium accounts for 99.985% of naturally occurring hydrogen and the deuterium makes up the remaining 0.015%. When hydrogen is mentioned, it is usually referred to as protium, the major isotope of hydrogen.

APPLICATIONS

As an isotope of hydrogen, deuterium exhibits the very same chemical properties as protium. On the other hand, deuterium has certain physical properties that are different from those of protium: it is twice as heavy as protium, which makes its bond to carbon or oxygen stronger than those attached to protium. Due to these unique properties, deuterium has been widely used in chemistry, biology, and physics.

The field of organic chemistry has benefited the most from the discovery of deuterium. One familiar example is the use of deuterated solvents such as deuteriochloroform ($CDCl_3$) in nuclear magnetic resonance (NMR) spectroscopy. The proton NMR (^1H NMR) spectrum of a sample provides valuable information about the structure of a molecule. In obtaining a proton NMR spectrum, a sample is typically dissolved in deuterated solvents such as deuteriochloroform. Obviously, deuterated solvent is required to clearly observe the signals arising from the analyte by obscuring the signal from the solvent.

Another application is the use of deuterium as a tracer in the study of reaction mechanism. With the use of deuterium-labeled compounds, organic chemists can conveniently follow the molecules to precisely figure out the reaction mechanism. An outstanding example can be found in a research paper published in 2010 by Professor Grubbs and coworkers at the California Institute of Technology. In studying the mechanism of ring-closing metathesis, the authors prepared a deuterium-labeled substrate (**1D2**) and subjected it to the ruthenium-catalyzed reaction (Scheme 1).[2] In addition to the expected product cyclopentene (**2**), two new compounds (**1D0, 1D4**) that differ only from the starting material diene (**1D2**) in isotopic composition could be detected by mass spectrometry.

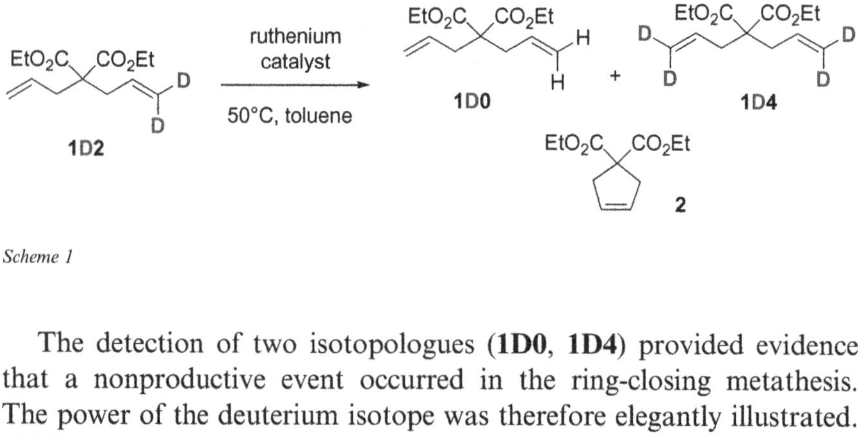

Scheme 1

The detection of two isotopologues (**1D0, 1D4**) provided evidence that a nonproductive event occurred in the ring-closing metathesis. The power of the deuterium isotope was therefore elegantly illustrated. Without deuterium, the study would not have been possible!

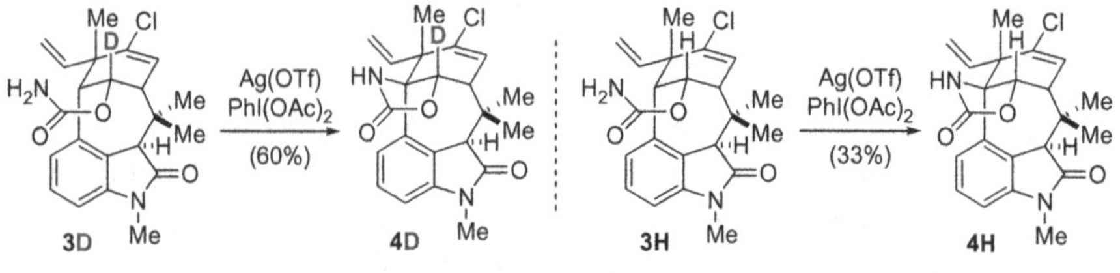

Scheme 2

The C–D bond reacts slower than the C–H bond. This particular effect is frequently exploited in synthetic organic chemistry. For example, Professor Neil Garg and coworkers at the University of California, Los Angeles, prepared a deuterium-labeled carbamate **3D** to accomplish a highly efficient C–H activation reaction in the total synthesis of (–)-*N*-methylwelwitindolinone C isonitrile (Scheme 2).[3] When subjected to the silver-promoted nitrene insertion reaction, the desired product was obtained from the carbamate **3D** twice as much as from the protium substrate **3H**.

The applications of deuterium-labeled compounds go beyond the areas of NMR spectroscopy, mechanistic studies, or total synthesis of natural products in organic chemistry. Recently, medicinal chemists at the pharmaceutical companies are testing the idea that simply substituting deuterium for protium in a currently approved drug could create a better drug. DeuteRx in Andover, MA, introduced in 2015 a deuterium-labeled thalidomide analog to explore the possibility of developing a single enantiomer drug for the treatment of multiple myeloma (Scheme 3).[4]

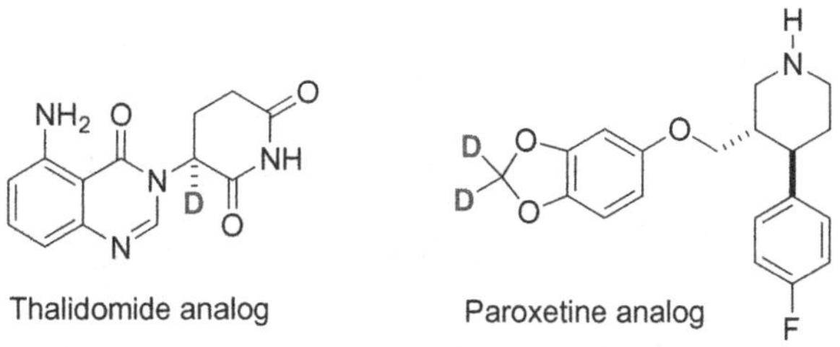

Thalidomide analog   Paroxetine analog

Scheme 3

Another example of deuterated drugs is by ConCert Pharmaceuticals of Lexington, MA, which reported very positive results of a Phase I clinical trial for a deuterium version of the antidepressant paroxetine, sold as Seroxat by GlaxoSmithKline.[5]

NO SINGLE BOOK IS FOUND

Considering that deuterium has had a tremendous impact on many areas of science, no single book exists that describes in detail how deuterium was discovered. Following a brief description of isotopes in

Chapter 1, Isotopes, the excitement and heroic efforts surrounding the discovery of deuterium are presented in Chapter 2, Deuterium. The stories are told in the narrative form extracted from the original research articles. A short note on how deuterium gas and deuterium oxide are manufactured is included as well. In Chapter 3, Deuterium-Labeled Compounds, basics of deuterium-labeled compounds such as their nomenclature and synthetic methods are described. In order to highlight the utility of deuterium, selected examples of applications in organic chemistry from earlier times to recent years are illustrated in Chapter 4, Applications in Organic Chemistry. Finally, Chapter 5, Applications in Medicinal Chemistry, outlines the biological effects of heavy water and the recent progress in the development of deuterated drugs.

This book would serve as an introductory reference on the history of deuterium and its applications in organic chemistry. I hope this book will be of use to those who are curious about deuterium.

REFERENCES

1. Rigden JS. *Hydrogen: the essential element.* Cambridge, MA: Harvard University Press; 2002.
2. Stewart IC, Keitz BK, Kuhn KM, Thomas RM, Grubbs RH. *J Am Chem Soc* 2010;**132**:8534.
3. Quasdorf KW, Huters AD, Lodewyk MW, Tantillo DJ, Garg NK. *J Am Chem Soc* 2012;**134**:1396.
4. Jacques V, Czarnik AW, Judge TM, Van der Ploeg LHT, DeWitt SH. *Proc Natl Acad Sci* 2015;**112**:E1471.
5. Uttamsingh V, Gallegos R, Liu JF, Harbeson SL, Bridson GW, Cheng C, Wells DS, Graham P, Zelle R, Tung R. *J Pharmacol Exp Ther* 2015;**354**:43.

CHAPTER 1

Isotopes

1.1 DEFINITION

Isotopes are a group of chemical elements that have the same number of protons, but have a different number of neutrons.[1] Isotopes thus have a different atomic mass, but maintain the same chemical characteristics. Nearly all the chemical elements that make up our material world occur in different isotopic forms. In fact, 83 of the most abundant elements have one or more isotopes composed of atoms with different atomic masses. Some familiar examples are chlorine (Cl), bromine (Br), carbon (C), and oxygen (O): chlorine has two stable isotopes of masses of 35 and 37; bromine has two isotopes of masses of 79 and 81; carbon has two stable isotopes of masses of 12 and 13; and oxygen has three stable isotopes of masses of 16, 17, and 18.

It was Frederick Soddy, who first proposed the word "isotopes" in 1913 in a paper published in *Nature*.[2]

> *So far as I personally am concerned, this has resulted in a great clarification of my ideas, and it may be helpful to others, though no doubt there is little originality in it. The same algebraic sum of the positive and negative charges in the nucleus, when the arithmetical sum is different, gives what I call "isotopes" or "isotopic elements," because they occupy the same place in the periodic table. They are chemically identical and save only as regards the relatively few physical properties, which depend upon atomic mass directly.*
>
> ***Reprinted with permission from Macmillan Publishers Ltd: Soddy, F.*** **Nature, *1913*, *92, 399. Copyright 1913.***

Deuterium. DOI: http://dx.doi.org/10.1016/B978-0-12-811040-9.00001-1

Soddy received the 1921 Nobel Prize in chemistry for his contributions to our knowledge of the chemistry of radioactive substances and his investigations into the origin and nature of isotopes:[3]

Soddy was born in Eastbourne, England, on September 2, 1877. He studied at Eastbourne College and the University College of Wales, Aberystwyth. In 1895, he obtained a scholarship at Merton College, Oxford, from which he graduated in 1898 with first class honors in chemistry. After 2 years of research at Oxford, he became a demonstrator in chemistry at McGill University in Montreal. At McGill, he worked on radioactivity with British physicist Sir Ernest Rutherford. Together they published a series of papers on radioactivity and concluded that it was a phenomenon involving atomic disintegration with the formation of new kinds of matter. In 1903, Soddy left Canada to work at University College London with Scottish chemist Sir William Ramsay. From 1904 to 1914, Soddy served as a lecturer at the University of Glasgow, Scotland. During this period, he evolved the so-called "Displacement Law," namely that emission of an alpha particle from an element causes that element to move back two places in the Periodic Table. In 1908, he married Winifred Beilby. The couple had no children. In 1919, he became Lee's Professor of Chemistry at Oxford University, a position he held until 1937 when he retired on the death of his wife. He died in Brighton, England, on September 22, 1956, at the age of 79.

1.2 ISOTOPES OF HYDROGEN

The hydrogen atom is the simplest of all atoms: it consists of a single proton and a single electron. In addition to the most common form of the hydrogen atom that is called protium, two other isotopes of hydrogen exist: deuterium and tritium. The atoms of deuterium (atomic symbol: D or ^{2}H) contain one proton, one electron, and one neutron, while those of tritium (atomic symbol: T or ^{3}H) contain one proton, one electron, and two neutrons. Whereas protium and deuterium are stable, tritium is not: it is radioactive. It is interesting to note that only the hydrogen isotopes have different names.

1.3 USES OF DEUTERIUM IN ORGANIC CHEMISTRY

Organic molecules that contain carbon–hydrogen bonds constantly undergo myriad reactions, in which reactants become products after going through a certain pathway. Organic chemists are very curious about the mechanism of the reaction, as a thorough understanding of the reaction mechanism not only provides the details of chemical change but also forms the foundation for invention of new reactions. Thus the elucidation of reaction mechanism is a rewarding process.

Among the many ways of studying reaction mechanism available to organic chemists, the use of deuterium is a very powerful tool especially when the reaction involves hydrogen.[4]

There are two properties that make deuterium so useful in organic chemistry. First, it is twice as heavy as protium (Table 1.1). When the hydrogen is replaced by deuterium, the resulting deuterium-labeled compound can be readily distinguished from the ordinary compound by mass spectra. Another advantage is taken of the heavy weight of deuterium in the study of deuterium kinetic isotope effect (DKIE). As the C–H bond breaks faster than the C–D bond, the measurement of the relative reaction rate gives a good idea about the reaction mechanism if that bond is involved in the reaction being investigated.

Second, deuterium has different magnetic properties than protium. Thus the C–H bonds of an organic compound can be detected by ^{1}H NMR, whereas the C–D bonds cannot.[5] The opposite is true with ^{2}H NMR spectroscopy. This unique feature makes it possible to follow deuterium attached to a specific carbon during the reaction. The combination of mass spectra with NMR spectroscopy thus makes it possible to employ deuterium as an isotopic tracer in an endeavor to solve the mechanism puzzle.

Deuterium is a rare isotope of hydrogen: there exists only one deuterium to about 6500 protiums found in nature. It is therefore no surprise that deuterium had only been discovered about 80 years ago. In the next chapter, we will learn how deuterium was discovered.

Table 1.1 Some Properties of Protium and Deuterium

Atom	H, protium	D, deuterium
Natural abundance	99.985%	0.015%
Atomic mass	1.008	2.014
Nuclear spin	1/2	1

REFERENCES

1. Krebs RE. *The history and use of our Earth's chemical elements*. Greenwood Press; 1998 p. 27–8.

2. Soddy F. *Nature* 1913;**92**:399.

3. NobelPrize.org. Frederick Soddy—biographical. *Chem Eng News* 2013;**December 2**:30–1.

4. Semenow DA, Roberts JD. *J Chem Educ* 1956;**33**:2.

5. Smith ICP, Mantsch HH. *Deuterium NMR spectroscopy*, vol. 191. ACS Symposium Series; **1982**; pp. 97–117 [Chapter 6].

CHAPTER 2

Deuterium

2.1 DISCOVERY

Deuterium was discovered by Harold C. Urey, professor of chemistry at Columbia University in the winter of 1931 right around the Thanksgiving holiday[1]:

> Harold Clayton Urey was born in Walkerton, IN, on April 29, 1893, as the son of the Rev. Samuel Clayton Urey and Cora Rebecca Reinoehl. In 1914, he entered the University of Montana to study both Zoology and Chemistry. It is those years at the University of Montana where he received his first inspiration for scientific work through personal relationship with his professors. In 1917, he obtained the Bachelor of Science degree in Zoology. In 1921, he entered the University of California at Berkeley to work under Professor Gilbert N. Lewis and was awarded the PhD degree in chemistry in 1923. He spent the following year in Copenhagen at Professor Niels Bohr's Institute for Theoretical Physics as an American-Scandinavian Foundation Fellow. After return to the United States, he joined the faculty at the Department of Chemistry at Johns Hopkins University. In 1929, he was appointed as associate professor of chemistry at Columbia University and became full professor in 1934. He was editor of the *Journal of Chemical Physics* during 1933–1940. Professor Urey received the Willard Gibbs Medal presented by American Chemical Society in 1934 for his work on the isotope of hydrogen. Later in

Deuterium. DOI: http://dx.doi.org/10.1016/B978-0-12-811040-9.00002-3

the same year, he received the Nobel Prize in chemistry. He was professor at the University of California, San Diego, since 1958 until his retirement. He married Frieda Daum in 1926. The couple had 3 daughters and 1 son, 4 grandchildren, and 10 great grandchildren in 1979. In his late years, he suffered heart attack, but recovered. He died on January 5, 1981.

When Francis W. Aston determined atomic weights of elements using mass spectrograph, he observed that many elements had isotopes. For example, neon has two (20 and 22), chlorine has two (35 and 37), bromine has two (79 and 81), and krypton has six (78, 80, 82, 83, 84, 86). Unlike these elements, Aston observed that hydrogen had only one isotope[2]:

From these figures, it is safe to conclude that hydrogen is a simple element and that its weight, determined with such constancy and accuracy by chemical methods, is the true mass of its atom.

In the 1920s, oxygen was the standard against which atomic weights of all chemical elements were determined. Oxygen had been considered a single atom until 1929 when Giauque and Johnston spotted an isotope of oxygen of mass 18.[3] Now that oxygen had two isotopes, the atomic masses based on the system $O = 16$ had to be revised. In analyzing the relative abundance of oxygen isotopes, Raymond T. Birge, professor of physics at the University of California, Berkeley, and Donald H. Menzel, professor of astrophysics at Lick Observatory, called for an investigation[4]:

Assuming that the abundance ratio is really 630 to 1 ($^{16}O/^{18}O$), it follows that atomic masses based on $^{16}O = 16$ should be 2.2 parts in 10^4 greater than those based on the chemical system $O = 16$. It is accordingly of importance to test Aston's mass spectrograph results on this new basis. Of the elements that permit an accurate comparison of the chemical and mass spectrograph results, there remains only hydrogen. The chemical value is 1.00777 ± 0.00002, as compared with Aston's 1.00778 ± 0.00015. Aston's value, reduced to the chemical scale, is 1.00756 on division by 1.00022 and the discrepancy appears to be outside the limits of error. It could be removed by postulating the existence of an isotope of hydrogen of mass 2, with a relative abundance $^{1}H/^{2}H = 4500$. It should be possible, although difficult, to detect such an isotope by means of band spectra.

Urey had been thinking of the hydrogen isotopes and became convinced of the presence of deuterium isotope after reading a paper by Birge and Menzel.[5] Using Debye's theory of the heat capacity of solid substances, Urey estimated the pressure of hydrogen gas over the solid hydrogen, and found out that the boiling points of the two hydrogens, protium and deuterium, should be markedly different. This calculation formed the basis of Urey's attempt to isolate deuterium through fractional distillation[6]:

The method of concentration—distillation of liquid hydrogen—came to me at lunch one day early in August 1931. I immediately discussed it with Dr. Murphy who was my research assistant. I had never considered the thermodynamic properties of solid hydrogen in detail and it took some time to straighten out all of the theoretical details, particularly the zero-point energy of a solid. I contacted Dr. Brickwedde at the Bureau of Standards to distill liquid hydrogen. Dr. Brickwedde, with whom I became acquainted at Johns Hopkins University, distilled 5- to 6-liter quantities of liquid hydrogen at the triple point of H_2, 14K at 53 mmHg, to a residue of 2 cm^3 of liquid, which was evaporated to glass flasks and sent to me.

Reprinted with permission from: Urey HC.* Ind Eng Chem *1934;* 26*:803. Copyright 1934 American Chemical Society.

To identify protium and deuterium, Urey and Murphy employed a spectroscopic method, using the Balmer series in the atomic spectrum of hydrogen. Recording the spectrum on a grating required a great deal of attention and time:

Upon receiving the distillation residue, Dr. Murphy and I went to work immediately and in one month did about four months' work. We did two ordinary days' work each day, labored Sundays and Thanksgiving Day as well. Mrs. Urey was a scientific widow for that month. To make sure that the spectral lines observed are real, we spent a whole month in and out of a dark room in the basement of the Physics Building, developing about 100 photographic plates. The situation increased our consumption of cigarettes about ten fold and made us quite unsuitable for human society.

Finally, Urey and Murphy obtained the atomic spectrum of hydrogen using two different batches of samples from Dr. Brickwedde, which did show the presence of the wavelengths of light calculated for a hydrogen atom of mass 2 (Table 2.1).[7]

The second sample of hydrogen evaporated near the triple point shows the spectral lines greatly enhanced, relative to the lines of ^1H, over both those of ordinary hydrogen and of the first sample. The $^2H\alpha$ line is resolved into

Table 2.1 Balmer Series Wavelengths: Calculated Versus Observed

Line	Hα	Hβ	Hγ	Hδ
Δλ calc.	1.793	1.326	1.185	1.119
Δλ obs.				
Ordinary hydrogen	–	1.346	1.206	1.145
First sample	–	1.330	1.199	1.103
Second sample	1.820	1.315	1.176	–

a doublet with a separation of about 0.16 Å in agreement with the observed separation of the $^1H\alpha$ *line. The relative abundance in ordinary hydrogen, judging from relative minimum exposure time is about 1:4000, or less, in agreement with Birge and Menzel's estimate. A similar estimate of the abundance in the second sample indicated a concentration of about 1 in 800. Thus an appreciable fractionation has been secured as expected from theory.*

Reprinted with permission from: Urey HC, Brickwedde FG, Murphy GM.* Phys Rev *1932;**39**:164. Copyright 1932 American Physical Society.

The discovery of deuterium was announced at the Thirty-Third Annual Meeting of the American Physical Society held at Tulane University in New Orleans in December of 1931.[8] The announcement did not come easy though, as there was an anecdote as to how Urey and Brickwedde were able to attend the meeting[9]:

After the discovery of deuterium, Urey faced a very practical problem reporting it—a problem to find funds for travel to scientific meetings. I received a phone call from Urey, telling me that it appeared he was not going to get funds to travel to the December 1931 American Physical Society meeting at Tulane University, where he planned to present a paper reporting the discovery of deuterium. He asked me if I could get travel funds and present the paper. For this, I had to see Lyman J. Briggs, assistant director of research and testing at the Bureau of Standards. Briggs, soon to be named NBS director, was an understanding and considerate physicist, who on learning of the work to be reported, made funds available for my travel. In the meantime, Bergen Davis, a prominent physicist at Columbia University, heard of Urey's problem and went to see Columbia President Nicholas Murray Butler, who made funds available for Urey's travel. So we both went to Tulane for the APS meeting, and Urey presented the ten-minute paper.

Reprinted with permission from AIP publishing: Brickwedde FG.* Phys Today *1982;**35**:34.

The existence of deuterium was further confirmed by mass spectrometry by Professor Bleakney at Princeton University in January of 1932.[10]

Later Urey recalled the time of his discovery[11]:

> *It was Thanksgiving Day and I was doing my work at Columbia. I knew immediately I'd hit on it, an important discovery. I hurried home and called to my wife, "Frieda, we have arrived!" When asked by a reporter, Urey said, "I'm not a genius and I'm not compared to Einstein. I am a great admirer of Einstein. My success came from hard work and luck." Urey stated, "My principal theme of research had been that all scientific work depends on the careful work of our predecessors and coworkers and that our rapid advance in the sciences is due largely to the freedom with which we publish the results of our own work."*

2.1.1 Atomic Weights of Protium and Deuterium

Urey proposed the name protium (from the Greek word *protos* meaning first) for the isotope of hydrogen having an atomic weight of 1, and deuterium (from the Greek word *deuteros* for second) that of atomic weight 2. In 1933, the exact atomic weights of protium and deuterium were 1.00778 and 2.01356, based upon a standard of ^{16}O having an exact atomic weight of 16.[12] The current numbers are 1.00783 and 2.0141 for protium and deuterium, respectively, which are based on the ^{12}C system. The basis for expressing the atomic weight values has been changed from ^{16}O to ^{12}C in 1961.

2.1.2 Deuteron Versus Deuton

Deuteron was adopted as the name for the nucleus of deuterium by the Committee on Nomenclature, Spelling, and Pronunciations.[13] The name "deuton" was objected by certain scientists in England, who believe it to be easily confused with the name neutron when used orally. Another candidate was diplon suggested by Lord Rutherford. At a symposium on heavy hydrogen, Dr. Ladenburg proposed during discussion the use of deuteron instead of deuton or diplon. He reported that this suggestion came from Professor Bohr and that he had brought it back to America via England where he had secured Lord Rutherford's willingness to adopt it. Finally, Dr. Urey himself agreed that a vote be taken and this vote resulted in approval for deuteron.[14]

2.2 DEUTERIUM GAS (D_2)

Deuterium gas (D_2, dideuterium) is a primary source of deuterium isotope for the synthesis of deuterium-labeled compounds. Deuterium gas can be conveniently generated by the reaction of D_2O with metals such as sodium, calcium turnings, or a mixture of calcium oxide and zinc (Scheme 2.1).[15] With a special setup, deuterium gas of 99% purity was obtained when sodium was used. Reaction with calcium produced deuterium that contained 90% D_2 and no more than 10% HD, whereas the use of a mixture of zinc and freshly dehydrated calcium oxide gave deuterium of 96% chemical purity in 90% chemical yield.

$$2\,Na + 2\,D_2O \longrightarrow 2\,NaOD + D_2$$

$$Ca + 2\,D_2O \xrightarrow{260^\circ C} Ca(OD)_2 + D_2$$

$$CaO + Zn + D_2O \xrightarrow{500^\circ C} CaZnO_2 + D_2$$

Scheme 2.1 Formation of deuterium.

In the presence of catalyst, deuterium adds to the alkenes and alkynes in the same manner as hydrogen does. For example, Rittenberg and Schoenheimer prepared a deuterium-labeled stearic acid in their study of intermediate metabolism. Thus reduction of methyl linoleate with deuterium in the presence of platinum oxide catalyst provided, after saponification, the expected product stearic acid-d_4 (Scheme 2.2, Eq. 1).[16] In an effort to determine physical data of organic compounds containing deuterium, Adams and McLean performed reduction of dimethylacetylene dicarboxylate in 1936 with deuterium (Scheme 2.2, Eq. 2).[17] Densities and melting points were determined for the ordinary succinate and deuterium-containing succinate. As expected, the deuterium-labeled succinate is indeed heavier than the ordinary succinate.

$$CH_3(CH_2)_4CH{=}CHCH_2CH{=}CH(CH_2)_7CO_2Me \xrightarrow[2.\ NaOH,H_2O]{1.\ D_2,\ PtO_2,\ pet.\ ether} CH_3(CH_2)_4CHDCHDCH_2CHDCHD(CH_2)_7CO_2H \quad (1)$$

$$MeO_2C{-}C{\equiv}C{-}CO_2Me \xrightarrow[EtOAc]{D_2,\ PtO_2} MeO_2C{-}CD_2{-}CD_2{-}CO_2Me \quad (2)$$

Compound	Density	mp (°C)
Dimethyl succinate-d_4	1.1450	17.0
Dimethyl succinate	1.1185	18.2

Scheme 2.2 Reactions of deuterium.

Wilkinson's catalyst, $RhCl(PPh_3)_3$, is a versatile homogeneous catalyst for hydrogenation. In the study of stereochemistry, Wilkinson and coworkers used deuterium gas to establish the stereochemistry of addition (Scheme 2.3).[18] Thus reduction of maleic acid with deuterium gave *meso*-1,2-dideuteriosuccinic acid proving that *cis*-addition of D_2 to the alkene occurred.

$$HO_2C\text{-}CH{=}CH\text{-}CO_2H \xrightarrow[C_6H_6/EtOH\ (1:1)]{D_2,\ RhCl(PPh_3)_3} HO_2C\text{-}CHD\text{-}CHD\text{-}CO_2H$$

Scheme 2.3 Wilkinson's catalyst.

A convenient way of generating D_2 gas in the laboratory from zinc metal and DCl in D_2O was recently reported. In this method, a two-chamber system was used: D_2 gas was generated in one chamber, and then it diffused into another chamber where reaction occurred in cyclopentyl methyl ether (CPME) (Scheme 2.4).[19]

$$Zn + 2\ DCl\ in\ D_2O \xrightarrow{rt} [ZnCl_2 + D_2] \xrightarrow[10\%\ Pd/C,\ CPME,\ rt]{substrate} Product$$

Substrate	Product	Results
Ph-CH=CH-CO_2Et	Ph-CHD-CHD-CO_2Et	99% y, 100% D
Ph-C≡C-CO_2Et	Ph-CD_2-CD_2-CO_2Et	94% y, >95% D
4-Br-C_6H_4-CO_2K (KO_2C, Br)	4-D-C_6H_4-CO_2H (HO_2C, D)	96% y, >96% D

Scheme 2.4 Generation and reaction of deuterium.

2.3 DEUTERIUM OXIDE (D_2O)

Deuterium oxide is a deuterium version of water, in which the protium of the usual water molecule is replaced by deuterium. That is why deuterium oxide is also called heavy water. Heavy water or deuterium oxide is used for many purposes. It is used as a moderator in nuclear reactors. For us organic chemists, heavy water is a prime source of deuterium when deuterium is incorporated into organic compounds. As such, it is good to know how heavy water was produced in the past.

Although heavy water is nowadays commercially available at a reasonable cost, the concentration of deuterium from the ordinary water in the early 1930s was a challenge. Three research groups were deeply involved in the production of heavy water: Professor Urey at Columbia University, Dr. Washburn at the National Bureau of Standards in Washington, DC, and Professor Lewis at the University of California, Berkeley. Electrolysis and fractional distillation were the methods of choice.

Within 6 months after discovery of deuterium in the hydrogen gas, Washburn of National Bureau of Standards and Urey of Columbia University reported in June 1932 a potentially useful way of concentration of the deuterium isotope by the electrolysis of water[20]:

> *Though the normal electrode potentials of the isotopes of all elements except hydrogen must be so nearly the same that no appreciable separation can be expected from any small differences, this may not be true in the case of hydrogen isotopes because the very large mass ratio. The small electrode difference combined with any difference in the diffusion of two species of ions through the cathode film would make possible a fractionation of the mixture probably with a resulting enrichment of the residual water with respect to deuterium, the species present at the smaller concentration. In that case it is obvious that a systematic fractionation by electrolysis should lead to two final fractions consisting of (1) pure ^{1}H and (2) the equilibrium mixture of ^{1}H and ^{2}H. On the basis of above reasoning, it appeared possible to concentrate the deuterium isotope by electrolysis of water, and such experiment was started at the Bureau of Standards on December 9, 1931. In the meantime, some of the electrolysis residues obtained from the commercial electrolysis of water for the production of oxygen were examined through photographs and found out that there was a very definite increase in the abundance of deuterium relative to protium in these residual solutions.*

Continuing the electrolysis work, Washburn and coworkers reported their findings on the water enriched in deuterium: Compared to a normal water of a density 1.000, the residual water had a slightly higher value of 1.0014. The freezing point and boiling point of the sample were 0.050 and 0.02°C higher than those of normal water.[21]

Encouraged by the report of Washburn and Urey, Professor Lewis and Macdonald at the University of California, Berkeley, carried out electrolysis experiment and they succeeded in obtaining quite pure heavy water[22]:

The fact is, the difference in properties between the two isotopes of hydrogen is so much greater than between any other pair of isotopes that in spite of the very small amount of deuterium present in ordinary hydrogen, several methods will lead to an almost complete separation of deuterium from protium. For this purpose, we therefore engaged at once in a process designed to reduce by electrolysis 10 liters of the water from the large electrolytic cell down to 1 mL or less. A current of 250 amperes was used for the electrolysis of water made up with one liter of 5M alkali from the large electrolytic cell and 9 liters of distilled water. At the end of five or six days, the electrolyte was reduced to one liter, ninety percent of which was then distilled from a copper kettle to afford 900 mL. The electrolysis/distillation cycles continued, and after fourth cycle half a mL of water was obtained. The density was 1.035, which meant 31.5% of all the hydrogen is deuterium. In the second run, concentrating from 20 liters to 0.5 mL, we obtained water of density of 1.073. Accordingly in this water 65.7% of the hydrogen is deuterium. By determining the density of the water in the various stage of concentration, we have attempted to determine the efficiency of the electrolytic separation: all the results agree with the assumption that five times as much hydrogen as deuterium is evolved. Based on this figure, if the water containing 65.7% deuterium were reduced by electrolysis to one-quarter of its volume, it would contain 99% of deuterium. Finally we can estimate the amount of the heavy hydrogen isotope in ordinary water: the measurements indicate that in Berkeley city water there is one part of D to about 6500 parts of H.

Reprinted with permission from AIP publishing: Lewis GN, Macdonald RT. J Chem Phys *1933;1:341.*

A pioneering work to demonstrate the possibility of obtaining heavy water by fractional distillation of water was carried out in 1933 by Washburn and Smith at the National Bureau of Standards at Washington, DC[23]:

It must be possible to fractionate water by distillation. To demonstrate this, 10 liters of water having a specific gravity of 1.000053 were distilled at atmospheric pressure in a still provided with a 35-foot rectifying column. An initial distillate of 200 mL and the final residue of 100 mL were compared as to density and found to differ by 64.9 ppm, the residue having increased by 53.3 ppm and the distillate having decreased by 13.2 ppm. Distillation fractionation is thus possible and should find practical application in combination with electrolysis fractionation.

Reprinted with permission from AIP publishing: Washburn EW, Smith ER. J Chem Phys *1933;1:426.*

That same year in 1933, Lewis and Cornish published their results on fractional distillation of water[24]:

It was apparent to the Berkeley group that the distillation could be attractive. This led to the decision to build a distillation plant for primary enrichment of deuterium by water distillation at 60°C. The laboratory plant consisted of two columns each 22 m high. The primary column was 30 cm in diameter; the second stage column was 5 cm in diameter. Both columns were filled with scrap aluminum turnings. The packing material performed poorly and was chosen because of limitations of funds. The plant went into operation on June 8, 1933. It had a feed to the primary tower of 1 L ordinary water/min. Enriched water from the first stage was used as a feed for the second stage column. The output of the distillation plant was further enriched by Lewis and Macdonald in their electrolytic plant. Although the water distillation plant did not perform in accord with expectations as a result of the failure of the packing, nevertheless, Lewis and Macdonald were soon producing 99% D_2O on the order of 1 g/wk. They used part of the material in their own researches. They were extremely generous in making the material available to scientists all over the world including Lawrence in Berkeley, Lauritsen at Cal Tech, and Rutherford at Cambridge in England.

Gilbert Newton Lewis, famous for the Lewis structures, was professor of chemistry at the University of California at Berkeley[25]:

Gilbert Newton Lewis was born in Weymouth, Massachusetts, on October 23, 1875. His family moved to near Lincoln, Nebraska, in 1884 and he spent two years at the University of Nebraska. He transferred to Harvard when his father became an executive at Merchants Trust Company in Boston. After earning his B.S. degree in 1896, he taught for a year at the Phillips Academy in Andover, MA. He obtained his Ph.D. degree under T. W. Richards in 1899. In 1905, Lewis accepted a staff position at MIT under Professor A. A. Noyes, where he remained until 1912. In 1912, Lewis was offered a Professorship and Chair of the College of Chemistry at the University of California at Berkeley. At Berkeley, one of his research interests was the study of isotopes in chemistry and physics. Lewis's research on isotopes is an example of his wide-ranging and prolific interests.

Lewis and Macdonald measured physical properties of D_2O, and some of the currently accepted values are listed in Table 2.2.[26]

Table 2.2 Some Properties of D_2O and H_2O

Properties	D_2O	H_2O
Molecular weight	20.03	18.02
Boiling point (°C)	101.72	100
Freezing point (°C)	3.82	0
Density at 20°C	1.1056	0.9982
Temperature of maximum density (°C)	11.6	4
Molecular volume at 20°C	18.2	18.05
Viscosity at 20°C (cP)	1.25	1.005

One of the outstanding properties of heavy water is its density: it is 10% larger than that of ordinary water. Urey rationalized this by reasoning that the ratio of the density of pure deuterium oxide and protium oxide should be the ratio of their molecular weights.[27] Another notable property is the freezing point: heavy water freezes at 4°C. It is interesting to see heavy water is 25% more viscous than the light water.

Heavy water exhibits unique properties in the biological systems. These features will be further discussed in Chapter 5, Applications in Medicinal Chemistry.

2.3.1 Current Way of Producing Heavy Water

Distillation is conceptually the simplest among many methods investigated. Normal water boils at 100°C, while heavy water boils at 101.7°C. However excessive tower volume is needed, requiring a large capital investment, that makes it a less attractive approach compared to other methods. Another way of producing heavy water is electrolysis, which is not optimal due to high-energy consumption. The very first commercial heavy water plant in the world was built in 1934 in Norway that used electrolysis to produce heavy water.

Considering the economy and separation efficiency, the most promising processes are based on chemical exchange reaction.[28] One of the very well-established processes is water–hydrogen sulfide process or Girdler sulfide process (GS process).[29] Upon request from the United States Atomic Energy Commission, the Girdler Corporation developed the hydrogen sulfide process in the 1950s to produce heavy water for the nuclear reactors at the Savannah River Site in South Carolina. So the process is called the Girdler sulfide process.

In the GS process, deuterium is transferred from a hydrogen sulfide molecule to a water molecule and vice versa.

$$H_2O + HDS \rightleftarrows HOD + H_2S$$

The extraction and enrichment steps used in the GS process at the Bruce Heavy eater Plant in Canada are presented in Fig. 2.1.

In the extraction section, water passes countercurrent to a circulating stream of hydrogen sulfide gas in three large sieve tray towers operating in parallel. The extraction tower has two distinct process temperature sections to achieve dynamic deuterium distribution: the top being cold at 30°C and the bottom being hot at 130°C. At low temperature, deuterium migrates to the water from the hydrogen gas. This slightly deuterium-enriched water then flows down to the hot section, where deuterium is transferred to the hydrogen gas. The continuous equilibrium results in the concentration of deuterium at the bottom of the cold tower and the top of the hot tower. At certain point, portions of those concentrated streams are withdrawn to the second stage for further enrichment. Enriched water at 20–30% D_2O concentration from the third stage is sent to a distillation unit to produce reactor grade heavy water, ie, 99.75% deuterium oxide. The GS process

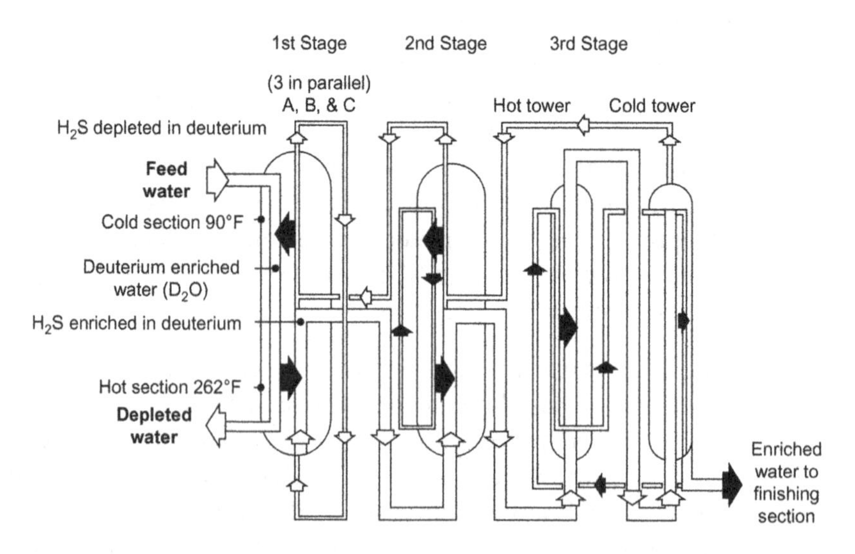

Figure 2.1 Diagram of the Girdler sulfide process. Reprinted with permission from: Davidson GD. Bruce heavy water performance. In: Rae HK, editor. *Separation of hydrogen isotopes*. ACS Symposium Series 68. Copyright 1978 American Chemical Society [chapter 2].

was successfully adopted in Canada to satisfy the demand for heavy water in CANDU (CANadian Deuterium Uranium) reactors.

Currently, deuterium oxide is commercially available to researchers from a variety of sources including Cambridge Isotope Laboratories, Inc., and the Sigma-Aldrich Corporation.

REFERENCES

1. NobelPrize.org. *Harold Clay Urey—bibliographical.*
2. Nobel lecture by Francis W. Aston; 1922.
3. Giauque WF, Johnston HL. *J Am Chem Soc* 1929;**51**:1436.
4. Birge RT, Menzel DH. *Phys Rev* 1931;**37**:1669.
5. Urey HC. *J Am Chem Soc* 1931;**53**:2872.
6. Urey HC. *Ind Eng Chem* 1934;**26**:803.
7. Urey HC, Brickwedde FG, Murphy GM. *Phys Rev* 1932;**39**:164.
8. Rigden JS. *Hydrogen: the essential element.* Cambridge, MA: Harvard University Press; 2002 [chapter 10]
9. Brickwedde FG. *Phys Today* 1982;**35**:34.
10. Bleakney W. *Phys Rev* 1932;**39**:536.
11. *Beaver County (Pennsylvania) Times*; May 21, 1979.
12. Urey HC. *Science* 1933;**78**:566.
13. Crane E. *J Ind Eng Chem News Edition* 1935;**13**:200.
14. For more stories Stuewer RH. *Am J Phys* 1986;**54**:206.
15. From sodium Mann WB, Newell WC. *Proc R Soc London A* 1937;**158**:397;
    From calcium Schiff HI, Steacie EWR. *Can J Chem* 1951;**29**:1;
    From calcium oxide and zinc Toby S, Schiff HI. *Can J Chem* 1956;**34**:1061.
16. Schoenheimer R, Rittenberg D. *J Biol Chem* 1935;**111**:163.
17. McLean A, Adams R. *J Am Chem Soc* 1936;**58**:804.
18. Osborn JA, Jardine FH, Young JF, Wilkinson G. *J Chem Soc A* 1966;1711.
19. Modvig A, Andersen TL, Taaning RH, Lindhardt AT, Skrydstrup T. *J Org Chem* 2014;**79**:5861.
20. Washburn EW, Urey HC. *Proc Natl Acad Sci* 1932;**18**:496.
21. Washburn EW, Smith ER, Frandsen M. *J Chem Phys* 1933;**1**:288.
22. Lewis GN, Macdonald RT. *J Chem Phys* 1933;**1**:341.
23. Washburn EW, Smith ER. *J Chem Phys* 1933;**1**:426.
24. Bigeleisen J. *J Chem Educ* 1984;**61**:108.
25. Harris HH, editor. *J Chem Ed*, 76. 1999. p. 1487.
26. Crespi HL, Katz JJ. In: Herber RH, editor. *Inorganic isotopic syntheses.* New York: W. A. Benjamin, Inc.; 1962. p. 43.

27. Urey HC. *Science* 1933;**78**:566.

28. Spindel W, Ishida T. *J Chem Educ* 1991;**68**:312.

29. Davidson GD. Chapter 2. Bruce heavy water performanceACS Symposium Series 68. In: Rae HK, editor. *Separation of hydrogen isotopes*. Washington, DC: American Chemical Society; 1978

CHAPTER 3

Deuterium-Labeled Compounds

3.1 NOMENCLATURE

Some 200 scientific papers dealing with deuterium were published in the 3-year interval between Urey's discovery and the time he received the Nobel Prize in 1934. The number of deuterium-labeled compounds increased dramatically due to the great interest in the uses of deuterium. It was therefore necessary to establish the nomenclature of those deuterium-labeled compounds. In 1934, a preliminary report on the nomenclature of the hydrogen isotopes and their compounds was published by the Committee on Nomenclature, Spelling, and Pronunciation of the American Chemical Society.[1]

In the naming of compounds containing one or more of deuterium, the compounds can be considered as derivatives of parent substances in which hydrogen (more precisely protium) has been replaced. With the attitude that compounds containing D are after all compounds of a

Deuterium. DOI: http://dx.doi.org/10.1016/B978-0-12-811040-9.00003-5

form of hydrogen, the committee has sought a system of nomenclature which would result in the least possible change from the established nomenclature for hydrogen compounds. The system is a modification of one proposed by Willis A. Boughton.[2] Other type of nomenclature is to use the prefix "deuterio" to indicate that the compounds contain one or more of deuterium (Table 3.1).

3.1.1 Isotopologues and Isotopomers

When a certain number of hydrogen in a molecule is replaced by a deuterium isotope, a new type of molecule is created that can be either isotopologue or isotopomer. Isotopologues are compounds with the same chemical structure that differ only in their isotope composition (Table 3.2).[3] On the other hand, isotopomers are isotopologues with the same number of deuterium.

Table 3.1 Names of Deuterium-Labeled Compounds

Compounds	Boughton Names	Deuterio Names
$CDCl_3$	Chloroform-d	Deuteriochloroform
CD_4	Methane-d_4	Tetradeuteriomethane
CH_3D	Methane-d	Deuteriomethane
C_6D_6	Benzene-d_6	Hexadeuteriobenzene
C_6H_5OD	Phenol-d	O-Deuteriophenol
$C_6H_5CO_2D$	Benzoic acid-d	Benzoic deuterioacid
ND_3	Ammonia-d_3	Trideuterioammonia
NH_2D	Ammonia-d	Deuterioammonia
D_2SO_4	Sulfuric acid-d_2	Sulfuric dideuterioacid
$HDSO_4$	Sulfuric acid-d	Sulfuric deuterioacid

Table 3.2 Isotopologues and Isotopomers

Isotopologues	Isotopomers
CD_3CH_2OH vs CH_3CD_2OH H_3C–CO–CD_3 vs D_3C–CO–CD_3 C_6H_5D vs C_6D_6	1-deuteriocyclohexanol (D, OH) vs cyclohexanol-OD (H, OD) 4-D-benzaldehyde (CHO, H) vs benzaldehyde-CDO (D, H)

3.1.2 Isotopic Steroisomers

Substitution of deuterium for hydrogen makes compounds optically active that would be otherwise achiral. Some examples are primary alcohols or amines with deuterium at C-1 carbon, or neopentane-d_6 (Scheme 3.1, Eq. 1). Meanwhile, an introduction of deuterium to an enantiomer produces a diastereomer (Scheme 3.1, Eq. 2).

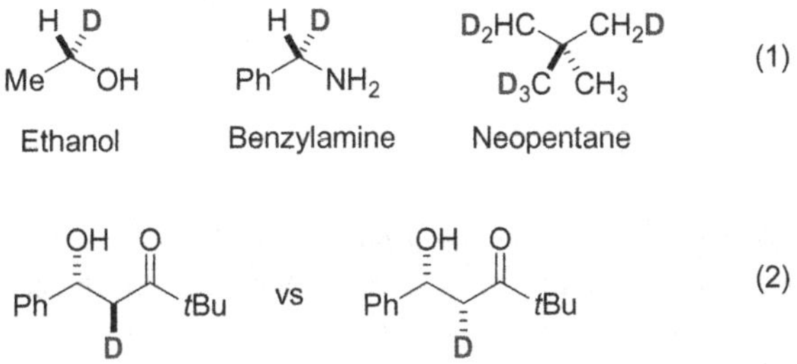

H D
Me OH
Ethanol
H D
Ph NH_2
Benzylamine
D_2HC CH_2D
D_3C CH_3
Neopentane
(1)

OH O
Ph tBu
D
vs
OH O
Ph tBu
D
(2)

Scheme 3.1 Optically active compounds.

Of course, geometric isomers are possible for ethylene with deuterium substitution (Scheme 3.2). Monosubstituted alkenes or 1,1-disubstituted alkenes can also have isomers, too.

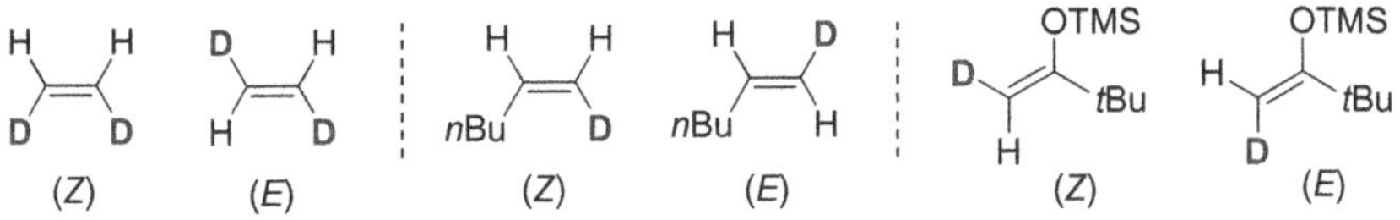

H H
D D
(*Z*)
D H
H D
(*E*)
H H
nBu D
(*Z*)
H D
nBu H
(*E*)
OTMS
D tBu
H
(*Z*)
OTMS
H tBu
D
(*E*)

Scheme 3.2 Geometric isomers.

3.2 SYNTHESIS OF ORGANIC COMPOUNDS

Deuterium-labeled compounds are prepared by a procedure known as hydrogen–deuterium (H/D) exchange reaction.[4] In the H/D exchange reaction, a substrate is mixed with either D_2 or D_2O along with the catalyst to affect the replacement of protium (H) by deuterium (D). Three types of catalysts are commonly used and these are acid, base, and a metal. In this section, preparations of several deuterium-labeled compounds are presented to showcase the chemistry involved in the H/D exchange reaction.

3.2.1 Benzene-d_6

Benzene-d_6 (C_6D_6), or hexadeuteriobenzene, was one of the first organic compounds prepared by H/D exchange reaction of the regular benzene (C_6H_6) with sulfuric acid-d_2 (D_2SO_4) and D_2O (Scheme 3.3).[5]

52% D_2SO_4 in D_2O
rt, 3 days
4 exchanges
99.8% D

Scheme 3.3

Sulfuric aicd-D_2 (D_2SO_4) was prepared from pure sulfur trioxide (SO_3) and deuterium oxide (D_2O) that contained 99.95 atom % deuterium. The concentration of the sulfuric acid was 52 mol%. Benzene was added to the sulfuric aicd-D_2, shaken at room temperature for 3–4 days. Benzene was then vacuum distilled to a fresh sample of sulfuric acid-D_2, and shaken again. After four repetitions, benzene-D_6 was obtained containing not less than 99.8 atom % deuterium.

The same method was applied to prepare naphthalene-d_8, which was used in the infrared studies of naphthalene and naphthalene-d_8 (Scheme 3.4).[6]

50% D_2SO_4 in D_2O
120°C, 50–100 h
4 exchanges
98% D

Scheme 3.4

Another way to prepare benzene-d_6 is via H/D exchange reaction with D_2O catalyzed by a platinum catalyst (Scheme 3.5).[7]

Pt/C
D_2O
110°C,12 h
4 exchanges
95% D

Scheme 3.5

A mixture of benzene (5 mL), deuterium oxide (10 mL), and platinum black (Pt/C, 0.3 g, prepared by reducing Adams' catalyst with deuterium) was heated in a sealed tube at 110°C for 12 hours. The benzene was distilled, which was subjected to three more cycles of H/D exchange reaction using platinum black and fresh deuterium oxide. The final benzene had 95.2% of C_6D_6 and 4% of C_6D_5H.

3.2.2 Mechanism of Pt-Catalyzed H/D Exchange Reaction

One plausible mechanism for the Pt-catalyzed H/D exchange reaction is as follows (Scheme 3.6).[8] Oxidative addition of deuterium oxide to the platinum metal yields Pt(D)(OD) species, which gives cationic platinum deuteride and deuteroxide. The platinum deuteride then adds to the benzene to produce an intermediate **A**, which upon deprotonation gives intermediate **B**. Reductive elimination of platinum regenerates the Pt(0) catalyst with the release of deuterated benzene-d_1.

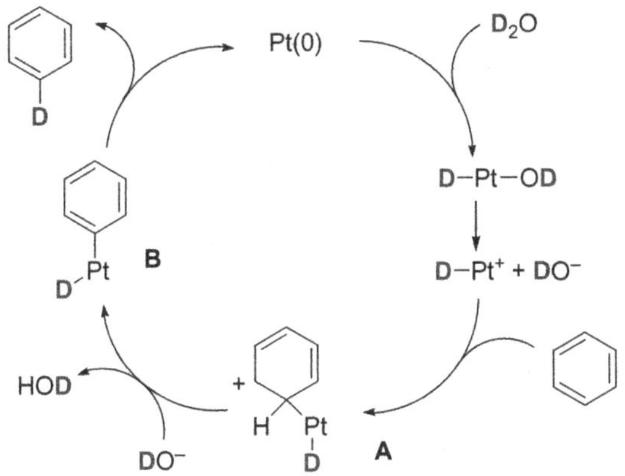

Pt(0) D₂O D–Pt–OD D–Pt⁺ + DO⁻ A HOD DO⁻ B D

Scheme 3.6 Mechanism.

The proposed platinum intermediate **B** indicates that the isotopic exchange belongs to a C–H bond activation/functionalization process.[9]

3.2.3 Acetone-d_6 and Other Ketones

Acidic C–H protons that are α- to the carbonyl group can undergo H/D exchange reaction under basic conditions. The first demonstration of a possible H/D exchange reaction of ketones with D_2O using a base catalyst was made on acetone in 1934, 2 years after the discovery of deuterium (Scheme 3.7).[10]

$$H_3CC(=O)CH_3 + D_2O\ (4.1\%\ D) \xrightarrow{K_2CO_3} H_3CC(=O)CH_2D\ (1.9\%\ D)$$

Scheme 3.7

When a solution of acetone (60 mL) in water (30 mL) containing 4.07% heavy water and a small amount of potassium carbonate (0.1 g) was warmed for 1–3 hours, an exchange of hydrogen atom occurred with the consequent introduction of deuterium into acetone. The exchange reaction was confirmed by the concomitant decrease in density of the water: the resulting water had 1.93% deuterium.

The authors predicted "the treatment of acetone with successively heavier portions of water will result in the practically complete replacement of protium by deuterium." Thirty years later, this was realized (Scheme 3.8).[11]

$$H_3CC(=O)CH_3 + D_2O\ (99.7\%\ D) \xrightarrow[\text{repeat 4 times}]{LiOD} D_3CC(=O)CD_3\ (\text{high D})$$

Scheme 3.8

The LiOD solution was prepared by reacting lithium wire with D_2O until LiOD began to precipitate. A small amount (0.4 mL) of a saturated solution of LiOD in D_2O was mixed with 50 mL of acetone and 100 mL of D_2O (99.7% D). The mixture was left standing for 30 minutes, after which the acetone was distilled. The H/D exchange reaction was repeated four more times to yield the highly pure acetone-d_6.

A variety of deuterated ketones can be prepared by base-catalyzed H/D exchange reaction. For example, 3-pentanone-2,2,4,4-d_4 was prepared by repeated exchanges with deuterium oxide in the presence of sodium carbonate (Scheme 3.9, Eq. 1).[12] Likewise, acetophenone-d_3 was readily prepared by H/D exchange reaction with D_2O in the presence of sodium hydroxide (Scheme 3.9, Eq. 2).[13] In this case, a single exchange reaction was sufficient to afford acetophenone of satisfactory deuterium content.

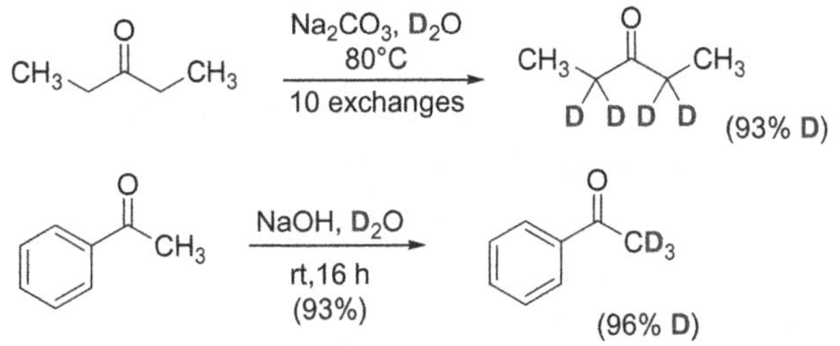

Scheme 3.9 H/D exchange reaction of ketones.

In studying the mechanism of the Favorskii rearrangement of α-halo ketones, deuterated cholestanone of high deuterium enrichment was prepared by three consecutive H/D exchange reactions using sodium deuteroxide in a mixture of heavy water and *p*-dioxane (Scheme 3.10).[14] The final deuterium incorporation was 98%.

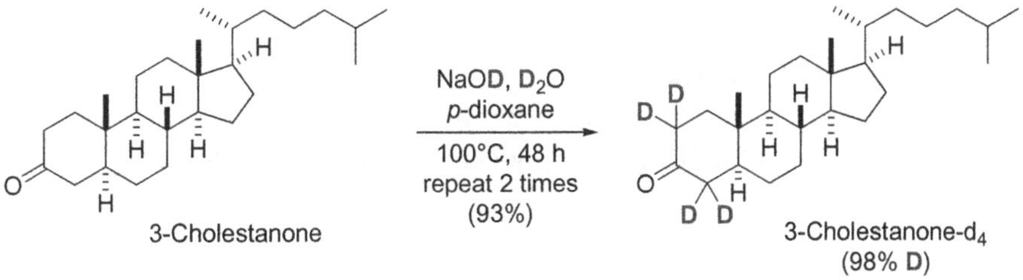

Scheme 3.10 3-Cholestanone-d_4.

3.2.4 Mechanism of Base-Catalyzed H/D Exchange Reaction

The effectiveness of a base-catalyzed H/D exchange reaction argues for a mechanism involving the postulated enolate of ketone (Scheme 3.11). The enolate form should exchange rapidly with deuterium of heavy water to produce monodeuterated ketone. A repeated H/D exchange reaction affords fully deuterated product.[15]

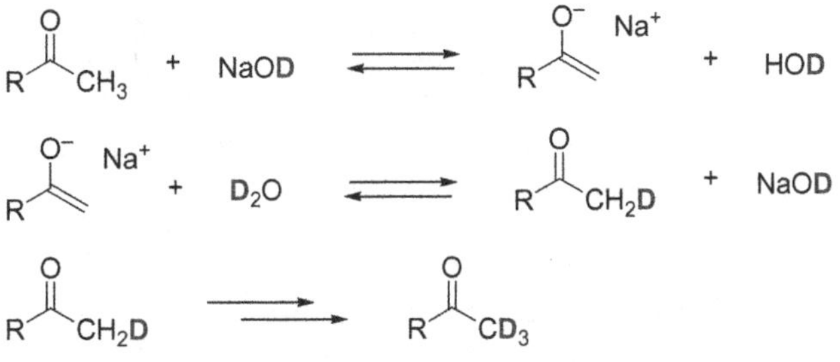

Scheme 3.11 Mechanism of H/D exchange reaction.

3.2.5 DMSO-d_6

The C–H protons of dimethyl sulfoxide ($pK_a = 35$) are acidic so that DMSO-d_6 can be prepared by H/D exchange reaction catalyzed by sodium deuteroxide (NaOD) (Scheme 3.12).[16]

Scheme 3.12

A solution of dimethyl sulfoxide in D_2O (DMSO:D_2O = 1:3 mole ratio) that contained 0.1 M NaOD was heated at 100°C for an hour before removal of water at 50 mmHg. The residue was further treated six times with D_2O, and finally distilled. The product showed no C–H protons by ^1H NMR and exhibited characteristic infrared frequencies reported for DMSO-d_6.

3.2.6 Chloroform-d ($CDCl_3$)

The first successful preparation of deuteriochloroform was reported in 1935 by reaction of chloral with sodium deuteroxide in heavy water (Scheme 3.13).[17]

Scheme 3.13 $CDCl_3$ from chloral.

Commercial chloral was redistilled to remove any residual ordinary water. The purified chloral (14.72 g, 0.1 mol) was then treated with deuterium oxide (5.47 g, 0.22 mol). The resulting chloral deuterate was added to the sodium deuteroxide (prepared from 2.2 g of sodium and 5 g of deuterium oxide) over a period of 5 h keeping the temperature below 5°C. After standing overnight, the reaction was completed by gentle warming of the reaction flask for 10 minutes. The layers were separated by centrifugation and the product was distilled and dried to afford 7.85 g of $CDCl_3$ in 68% yield. The boiling point of $CDCl_3$ was 0.5°C higher than that of ordinary chloroform, and the density of $CDCl_3$ was 1.5004 as compared with 1.4888 for ordinary chloroform.

Later in 1951, above synthesis of $CDCl_3$ was repeated and it was determined by infrared spectrometric and mass spectral analyses that the isotopic purity of $CDCl_3$ was not more than 96 percent.[18]

Having reasoned that the hydrogen of the chloral could be the protium source, a new route was devised (Scheme 3.14, Eq. 1). Trichloroacetophenone, which was obtained by careful chlorination of dichloroacetophenone, was treated with sodium deuteroxide at 0°C for 30 minutes. Although $CDCl_3$ was obtained in a marginal 30% yield, the isotopic purity of deuteriochloroform was greatly improved to 99.2% (0.8% $CHCl_3$) as determined by mass spectrum.

$$PhC(O)CCl_3 + NaOD \xrightarrow[\text{0°C, 0.5 h (30\%)}]{D_2O} CDCl_3 + PhC(O)ONa \quad (1)$$

$$Cl_3CC(O)CCl_3 + D_2O + \text{Pyridine} \xrightarrow[\text{(79\%)}]{\text{heat to distill}} CDCl_3 + CO_2 \quad (2)$$

Scheme 3.14 $CDCl_3$ via haloform reaction.

By utilizing a similar haloform reaction, a large-scale synthesis of $CDCl_3$ was reported in 1963 starting with hexachloroacetone (Scheme 3.14, Eq. 2):[19]

Hexachloroacetone (265 g, 1.0 mol) was mixed with heavy water (40 mL, 44 g, 2.2 mol) and pyridine (10 mL, 9.8 g, 0.012 mol). The contents were slowly heated to distill a mixture of $CDCl_3$, D_2O, and pyridine. The deuteriochloroform was then purified by a second distillation, dried over $CaSO_4$, and redistilled to afford 190 g (1.591 mol) of $CDCl_3$ in 79% yield.

Another way to the $CDCl_3$ synthesis is via the H/D exchange reaction of chloroform with D_2O under basic conditions (Scheme 3.15).[20]

$$CHCl_3 + D_2O \xrightarrow[\text{boiling water bath, 72 h}]{\text{0.1 N KOH (0.5 mL)}} CDCl_3$$

Scheme 3.15 $CDCl_3$ from $CHCl_3$.

Ordinary chloroform (20 mL) was mixed with 0.5 mL of 0.1 N KOH solution in water containing 4.02% deuterium. The reaction vessel was then sealed air free and heated in a boiling water bath for 72 hours. From the density measurements, the D-contents of the water and the resulting chloroform were 3.03% and 0.26%, respectively.

Based on the kinetic data, the authors found that the H/D exchange reaction was 90 times faster than the decomposition of chloroform under the basic conditions. The study was the first of its kind to demonstrate that deuteriochloroform could be made via an H/D exchange reaction of ordinary chloroform under basic conditions using D_2O.

Later in 1954, Hine and coworkers investigated the base-catalyzed H/D exchange reaction of chloroform in alkaline solution in D_2O and confirmed that the hydrogen atom in chloroform is about as reactive as that in acetone.[21]

3.3 $LiAlD_4$ AND $NaBD_4$

Reduction of carbonyl groups is an important reaction in organic chemistry. Two of the most frequently used reagents are lithium aluminum hydride ($LiAlH_4$) and sodium borohydride ($NaBH_4$).

Lithium aluminum deuteride ($LiAlD_4$), a deuterium version of $LiAlH_4$, can be prepared from lithium deuteride and aluminum bromide ($AlBr_3$) (Scheme 3.16).[22]

$$4\ LiD + AlBr_3 \xrightarrow[\text{reflux, 3–4 h}]{Et_2O} LiAlD_4$$

Scheme 3.16 Preparation of $LiAlD_4$.

A solution of aluminum bromide (267 g, 1.0 mol) in 750 mL of diethyl ether is added with cooling in an ice salt bath to the lithium deuteride (37 g, 4.13 mol) in 250 mL of diethyl ether in a 2 L flask. The mixture is heated to boiling and stirred for 3–4 hours. On cooling, lithium bromide settles, and a few grains of lithium deuteride float. The mixture is decanted to obtain a clear solution of lithium aluminum deuteride (40 g, 0.95 mol) in 95% yield.

Sodium borodeuteride ($NaBD_4$) can also be conveniently prepared in two steps starting from trimethylamineborane (Scheme 3.17).[23] The first step of the synthesis is the acid-catalyzed H/D exchange reaction of boron hydrogen for deuterium in heavy water. The H/D exchange reaction was conveniently followed by IR spectroscopy: the broad band at 2300 cm^{-1} of the B–H bond weakened and at the same time a broad doublet at 1740 cm^{-1} of the B–D bond intensified as the reaction progressed. The second step of the synthesis is the reaction of the trimethylamineborane-d_3 with alcohol-free sodium methoxide to form sodium borodeuteride.

$$Me_3N{\bullet}BH_3 \xrightarrow[\substack{rt,\ 24\ h \\ \text{repeat 9 times} \\ (60\%)}]{\substack{0.5\ N\ D_2SO_2 \\ D_2O}} Me_3N{\bullet}BD_3 \xrightarrow[\substack{150°C,\ 4\ h \\ (73\%)}]{\substack{NaOMe \\ (EtOCH_2CH_2)_2O}} NaBD_4$$

Scheme 3.17 Synthesis of $NaBD_4$.

A solution of trimethylamineborane (400 g, 3.48 mol) in 4 L of anhydrous diethyl ether was vigorously stirred with 500 mL of 0.5 N D_2SO_4 in D_2O for 24 h, after which equilibrium of H/D exchange had been achieved. The layers were separated, 500 mL of fresh 0.5 N D_2SO_4 in D_2O were added, stirred for 24 h. After a total of 10 H/D exchange reactions, the ether layer was washed once with 200 mL of D_2O, dried over Na_2CO_3, filtered, and concentrated to afford 240 g (60% yield) of trimethylamineborane-d_3 as white crystals. A mixture of trimethylamineborane-d_3 (190 g, 1.61 mol) and NaOMe (65 g, 1.20 mol) in 400 mL of diglyme was heated at 150°C under N_2 for 4 h until trimethylamine evolution ceased. The insoluble material was collected via hot filtration, dried to yield 51 g of crude product. Recrystallization from n-propylamine *gave 37 g (73% yield) of sodium borodeuteride of 97% chemical purity and of 97–98% isotopic purity (deuterium content).*

REFERENCES

1. Crane E. *J Sci* 1934;**80**:86.
   Crane EJ. *Ind Eng Chem News Ed* 1935;**13**:200.
2. Boughton WA. *Science* 1934;**79**:159.
3. McNaught AD, Wilkinson A. *IUPAC compendium of chemical technology*. 2nd ed. 1997.
4. Thomas AF. *Deuterium labeling in organic chemistry*. New York: Appleton-Century-Crofts; 1971.
   Atzrodt J, Derdau V, Fey T, Zimmermann J. *Angew Chem Int Ed* 2007;**46**:7744.
5. Ingold CK, Raisin CG, Wilson CL. *J Chem Soc* 1936;915.
6. Oerson WB, Pimentel GC, Schnepp O. *J Chem Phys* 1955;**23**:230.
7. Leitch LC. *Can J Chem* 1954;**32**:813.
8. Yamamoto M, Oshima K, Matsubara S. *Org Lett* 2004;**6**:5015.
9. Labinger JA, Bercaw JE. *Nature* 2002;**417**:507.
   Hartwig JF. *J Am Chem Soc* 2016;**138**:2.
10. Halford JO, Anderson LC, Bates JR. *J Am Chem Soc* 1934;**56**:491.
    Bonhoeffer KF, Klar R. *Naturwissenschaften* 1934;**22**:45.
11. Paulsen PJ, Coole WD. *Anal Chem* 1963;**35**:1560.

12. Leitch LC, Morse AT. *Can J Chem* 1953;**31**:785.
13. Horino Y, Kimura M, Tanaka S, Okajima T, Tamaru Y. *Eur J Org* 2003;**9**:2419.
14. Nace HR, Olsen BA. *J Org Chem* 1967;**32**:3438.
15. Kawazoe Y, Ohnishi M. *Chem Pharm Bull* 1966;**14**:1413.
16. Buncel E, Symons EA, Zabel AWJ. *J Chem Soc Chem Commun* 1965;173.
17. Breuer FW. *J Am Chem Soc* 1935;**57**:2236.
18. Boyer WM, Bernstein RB, Brown TL, Dibeler VH. *J Am Chem Soc* 1951;**73**:770.
19. Paulsen PJ, Coole WD. *Anal Chem* 1963;**35**:1560.
20. Horiuti J, Sakamoto Y. *Bull Chem Soc Jpn* 1936;**11**:627.
21. Hine J, Peek Jr RC, Oakes BD. *J Am Chem Soc* 1954;**76**:827.
22. Wiberg E, Schmidt M. *Z Naturforsch* 1952;**7b**:59.
    Corval M, Bengsch E. *Bull Soc Chim Fr* 1967;2295.
23. Atkinson JG, MacDonald DW, Stuart RS, Tremaine PH. *Can J Chem* 1967;**45**:2583.

CHAPTER 4

Applications in Organic Chemistry

Deuterium. DOI: http://dx.doi.org/10.1016/B978-0-12-811040-9.00004-7

4.1 BACKGROUND

Gilbert N. Lewis expressed his excitement about the uses of deuterium in organic chemistry[1]:

> *The isotope of hydrogen is, beyond all others, interesting to chemists. I believe that it will be so different from common hydrogen that it will be regarded almost as a new element. If this is true, the organic chemistry of compounds containing the heavy isotope of hydrogen will be a fascinating study.*

With deuterium, organic chemists can do a lot of amazing experiments. One of them is the study of kinetic isotope effect (KIE).

4.1.1 Kinetic Isotope Effect

Deuterium KIEs are widely used in physical organic chemistry to determine which step of the reaction is the rate-determining step. KIE arises because of the reactivity difference of the hydrogen and deuterium. The hydrogen reacts faster than the deuterium isotope.

It was Urey who first observed KIE in 1934, in the synthesis of methane-d_4 (Scheme 4.1).[2] Ordinary water reacted much faster than heavy water by more than 20 times.

$$Al_4C_3 + 12\,H_2O \xrightarrow{k_H} 4\,Al(OH)_3 + 3\,CH_4$$

$$Al_4C_3 + 12\,D_2O \xrightarrow{k_D} 4\,Al(OD)_3 + 3\,CD_4$$

$$\frac{k_H}{k_D} = 23$$

Scheme 4.1 Reactions of aluminum carbide with water.

> *Carefully prepared aluminum carbide was placed in a sealed container fitted with a small reflux condenser and ordinary water was added at low temperature. When the mixture was allowed to warm to room temperature, gas bubbles formed in the course of 10 or 15 minutes. The reaction mixture was then heated to 80°C and the methane gas was collected. It required on the average 2 minutes to collect 100 mL of methane. With the same apparatus and aluminum carbide from the same batch, no reaction took place at room temperature with heavy water. Some bubbles of methane formed at 65°C. When the mixture was heated to 80°C, it now required 46.5 minutes to collect the first 100 mL of methane gas. This gives a ratio of the velocities for the two reactions of approximately 23. Thus an enormous difference in the velocities was observed.*
>
> **Reprinted with permission from AIP publishing: Urey HC, Price D. J Chem Phys *1934;2:300.***

A series of simple but elegant experiments can be conducted in the laboratory to visualize the KIE by reacting metals with water. For example, sodium and calcium carbide react with water to produce hydrogen and acetylene gas, respectively (Scheme 4.2).[3] It is clear that both sodium and calcium carbide react faster with ordinary water than with heavy water.

$$2\,Na + 2\,H_2O \longrightarrow 2\,NaOH + H_2$$

$$2\,Na + 2\,D_2O \longrightarrow 2\,NaOD + D_2$$

$$CaC_2 + 2\,H_2O \longrightarrow Ca(OH)_2 + H{-}C{\equiv}C{-}H$$

$$CaC_2 + 2\,D_2O \longrightarrow Ca(OD)_2 + D{-}C{\equiv}C{-}D$$

Scheme 4.2 Reactions of sodium and calcium carbide with water.

> Reactions with sodium: *In separate 18 × 150 mm test tubes are placed 5 mL of* H_2O *and 5 mL of* D_2O*. Two equal pea-sized lumps of sodium are simultaneously dropped into the test tubes. The hydrogen gas released from the test tube with* H_2O *bursts into flames and a mild explosion usually occurs. The reaction in* D_2O*, while vigorous, is not as violent as the reaction in* H_2O*. The sodium lump scoots about on the surface of the heavy water (*D_2O*) without flame or explosion!*
>
> Reactions with calcium carbide: *Equal amounts (0.25 g) of calcium carbide are placed in 18 × 150 mm test tubes. Next 4 mL of* H_2O *and 4 mL of* D_2O *are added simultaneously to the calcium carbide. The tube with* H_2O *froths more vigorously than the tube with* D_2O*. At this point, both tubes are ignited. In the tube containing* D_2O*, soot descends and adheres to the inner surface of the tube. Such is not observed in the tube with* H_2O*. This is because the acetylene is liberated faster from the tube with* H_2O *than the tube with* D_2O*.*
>
> ***Reprinted with permission from: Binder DA, Ellason R. J Chem Ed 1986;63:536.***
> ***Copyright 1986 American Chemical Society.***

To measure the relative reactivity of O–H and O–D bonds, Wilson and coworkers dissolved the metals in a known amount of water. The hydrogen gas was collected and analyzed for the isotopic composition (Table 4.1).[4]

The isotopic separation coefficient, α, is an expression of the relative reactivity of O–H bond over O–D bond, for example, ordinary water reacts 2.82 times faster than heavy water with sodium.

A subtle difference in reactivity between the hydrogen and deuterium was observed in the metalation of 2-methyl-4-phenylthiazole (Scheme 4.3).[5] When the thiazole **1** in tetrahydrofuran was treated with *n*-butyllithium at −78°C, and subsequently quenched with methyl iodide, 2,5-dimethyl-4-phenylthiazole **2** was formed along with 2-ethyl-4-phenylthiazole **3** in a 19:1 ratio. The ratio abruptly changed when the 5-deuterio derivative **1D** was subjected to the same conditions. The two products (**2** and **3D**) were formed in equal amounts.

n-BuLi, −78°C; then MeI: **1** → **2** + **3** (94.8:5.2)

n-BuLi, −78°C; then MeI: **1D** → **2** + **3D** (49.3:50.7)

Scheme 4.3

Table 4.1 Relative Rate of Reaction

Metal	Aqueous Medium	α
Li	$H_2O \rightarrow$ aq. LiOH	1.49
Na	$H_2O \rightarrow$ aq. NaOH	2.82
K	$H_2O \rightarrow$ aq. KOH	1.95
CaC_2	$H_2O \rightarrow$ aq. $Ca(OH)_2 + C_2H_2$	1.50

It was calculated from the data that the C–H bond reacted with *n*-butyllithium 19 times faster than the C–D bond at −78°C (Scheme 4.4).

Ph, N, H, S, Me — nBuLi, −78°C, k_H → Ph, N, Li, S, Me

Ph, N, D, S, Me — nBuLi, −78°C, k_D → Ph, N, Li, S, Me

$\frac{k_H}{k_D} = 18.8$

Scheme 4.4 Kinetic isotope effect.

Why is there such a difference in the reactivity between protium and deuterium?

When a bond to hydrogen or deuterium is broken in the rate-determining step of a reaction, the rate constant, k_H, for the reaction of the hydrogen compound exceeds the constant, k_D, for the same reaction of the corresponding deuterium compound. Lower reactivity of bonds to deuterium as compared to the corresponding bonds to hydrogen is due to the difference in zero-point energy (Fig. 4.1).[6] The heavy isotope deuterium has a lower zero-point energy than the lighter isotope hydrogen, which is due to the effect of the difference in mass on the stretching frequencies.

In transition state theory of reaction, the reaction goes through an activated complex where the bond undergoing reaction is relatively weak compared to the bond in the reactants (Fig. 4.2). The weak bond in the activated complex reflects a low force constant resulting in a small difference in zero-point energy for this bond in the activated complex. In mathematical terms: $E_0^H - E_0^D \gg [E_0^H]^\ddagger - [E_0^D]^\ddagger$. Thus the difference in zero-point energy in the reactants will result in a difference in the height of the potential-energy barrier for reaction.[7]

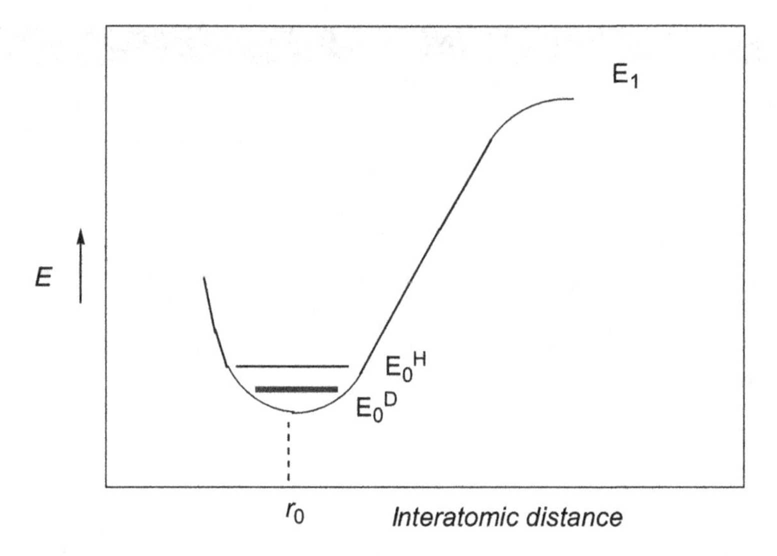

Figure 4.1 Morse curve relating potential energy and atomic distance. Source: Reprinted with permission from: Wiberg KB. *Chem Rev* 1955;**55**:713. Copyright 1955 American Chemical Society.

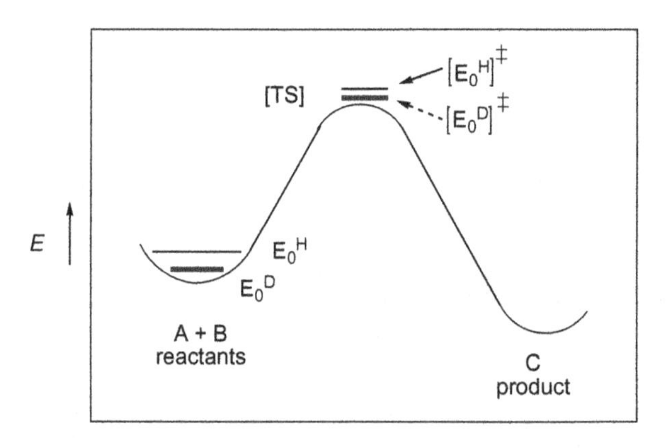

Figure 4.2 Reaction coordinate for a single step reaction.

The maximum isotope effect is observed when the reaction goes through a transition state where the protiated and deuterated activated complexes have the same energy. In this case, the loss of zero-point energy in the reactants will be the sole contributor, and the relative rate can be calculated by the following equation:

$$\frac{k_{\mathrm{H}}}{k_D} = e^{(h\nu_{\mathrm{H}} - h\nu_{\mathrm{D}})/2\mathrm{RT}}$$

Table 4.2 Magnitude of Isotope Effects

Bond	k_H/k_D at 25°C[a]	ΔE_0
C–H	6.9 (8.3)	~1.15 kcal/mol
O–H	10.6 (12.6)	~1.4 kcal/mol
N–H	8.5 (10.3)	~1.27 kcal/mol

[a]*Numbers in parentheses are at 0°C.*

Using the above equation, the maximum values of the deuterium isotope effect are as follows (Table 4.2). Note that the isotope effect becomes larger as the temperature goes lower.

There are two kinds of isotope effects: primary and secondary. Primary isotope effect is observed in a reaction where the C–H (C–D) bond breaks.[8] Primary isotope effects can provide two very useful pieces of information about a reaction mechanism. First, the existence of a substantial isotope effect, ie, if k_H/k_D is 2 or more, is strong evidence that the bond to the deuterium-substituted hydrogen atom is being broken in the rate-determining step. Second, the magnitude of the isotope effect provides a qualitative indication of where the transition state lies with respect to product and reactant. A relatively low primary isotope effect implies that the bond to the hydrogen is either slightly or nearly completely broken at the transition state. That is, the transition state must occur quite close to reactant or to product. An isotope effect near the theoretical maximum is good evidence that the transition state involves strong bonding of the hydrogen to both its new and old bonding partner. Secondary isotope effects are observed when the substituted hydrogen atom is not broken in the reaction and are smaller than the primary isotope effects.

4.1.2 Deuterium Tracer Study

Urey predicted that the use of deuterium would be valuable in organic chemistry by tagging the compounds[9]:

> *In many cases it should be possible to introduce a deuterium atom in some part of a complicated organic compound and then follow this atom in subsequent chemical reactions. It is, in fact, a "tagged" atom that can be introduced into chemical compounds at one point and later identified after rather extensive changes have taken place.*

Deuterium tracer study is another favorite application used by organic chemists in studying reaction mechanism. One elegant example is provided by Kochi and coworkers at Keio University in Japan, who employed a deuterium tracer strategy in their study of cyclization reaction (Scheme 4.5).[10]

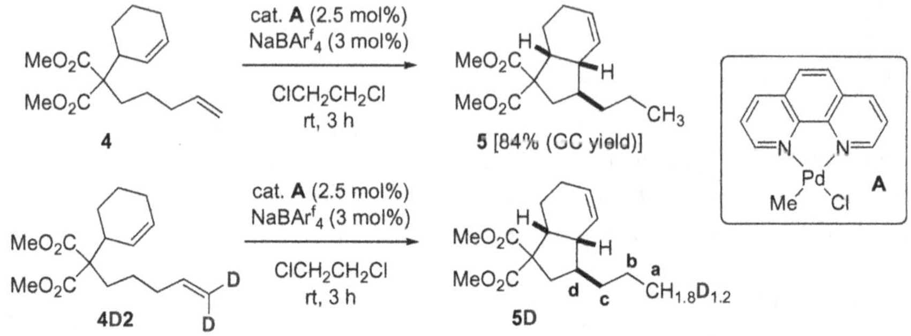

Scheme 4.5 Deuterium as a tracer.

When the diene **4** was treated with palladium catalyst **A**, a cyclization product **5** was obtained in 84% GC yield. In an attempt to establish that the reaction occurs through a chain-walking mechanism, a deuterium-labeled substrate **4D2** was prepared and subjected to the same conditions. As expected, the product **5D** has deuterium distributed through four carbons $C_a - C_d$. The results confirm that the cyclization does indeed proceed via a chain-walking mechanism.

NMR spectroscopy is a very useful analytical tool to keep track of the molecules tagged with deuterium. As deuterium is not detected in proton NMR (^{1}H NMR), the tagging or labeling of an organic molecule with deuterium is readily noticed by the absence of a proton signal at the site of deuterium substitution. The presence of deuterium is on the other hand confirmed by deuterium NMR (^{2}H NMR).

For example, the ^{1}H NMR spectrum of deuterium-labeled diene **4D2** shows that the vinylic CH_2 hydrogens are missing at 5 ppm (Fig. 4.3). The deuterium substitution was, however, confirmed by ^{2}H NMR (Fig. 4.4). The ^{2}H NMR spectrum of **5D** revealed a very interesting pattern: some of the deuterium atoms originally attached at the terminal vinylic carbon have moved along the carbon chain spanning C_a through C_d (Fig. 4.4).

As deuterium is twice as heavy as protium, mass spectrometry is a great way to detect the mass increase in deuterium-labeled compounds due to the introduction of deuterium to ordinary organic compounds.

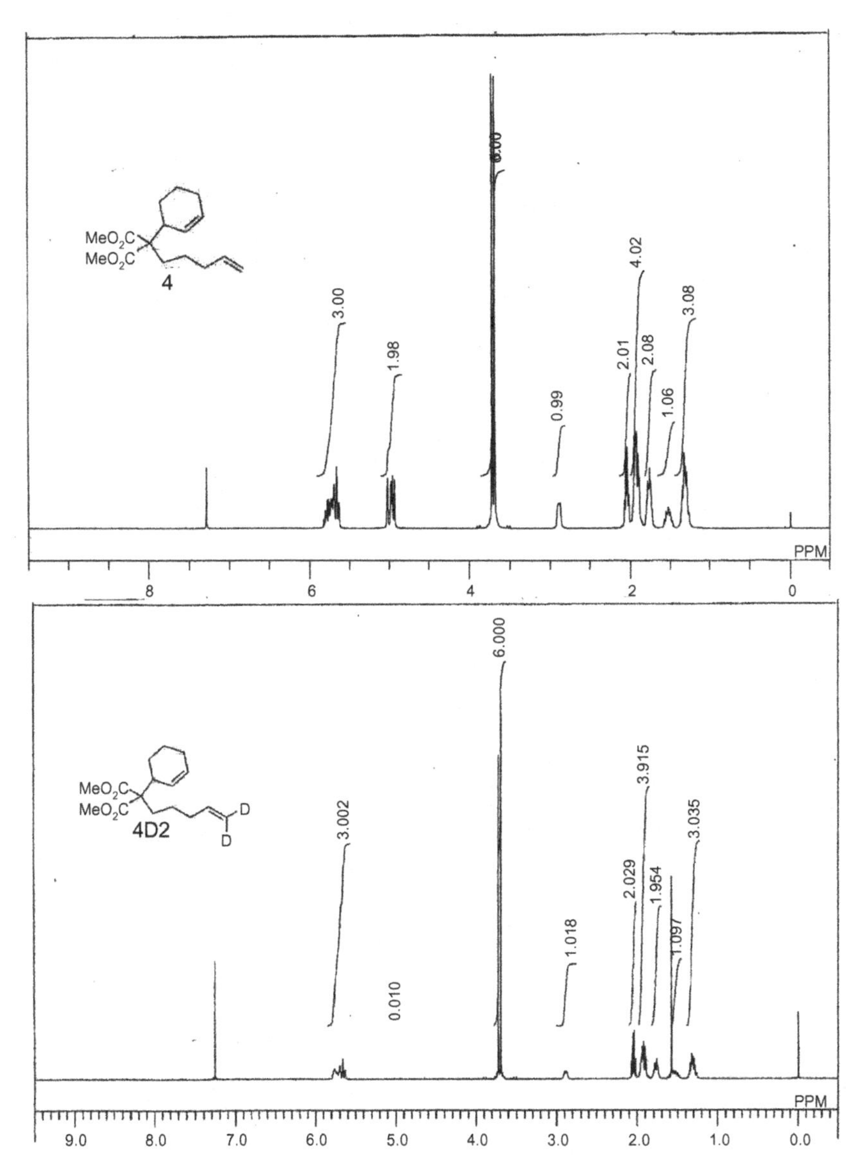

Figure 4.3 ¹H NMR spectra in $CDCl_3$ of ordinary diene and deuterium-labeled diene. Source: Reprinted with permission from: Kochi T, Hamasaki T, Aoyama Y, Kawasaki J, Kakiuchi F. *J Am Chem Soc* 2012;**134**:16544. Copyright 2012 American Chemical Society.

In order to demonstrate the mechanism of hydride reduction of benzaldehyde, Robinson and De Jesus of Union College in New York performed two parallel experiments: one was with $NaBH_4$ and the other with $NaBD_4$ (Scheme 4.6).[11]

+ $NaBH_4$

NaOMe, MeOH
rt, 5 min; then
5% HCl, H_2O

+ $NaBD_4$

same as above

Scheme 4.6 $NaBH_4$ versus $NaBD_4$.

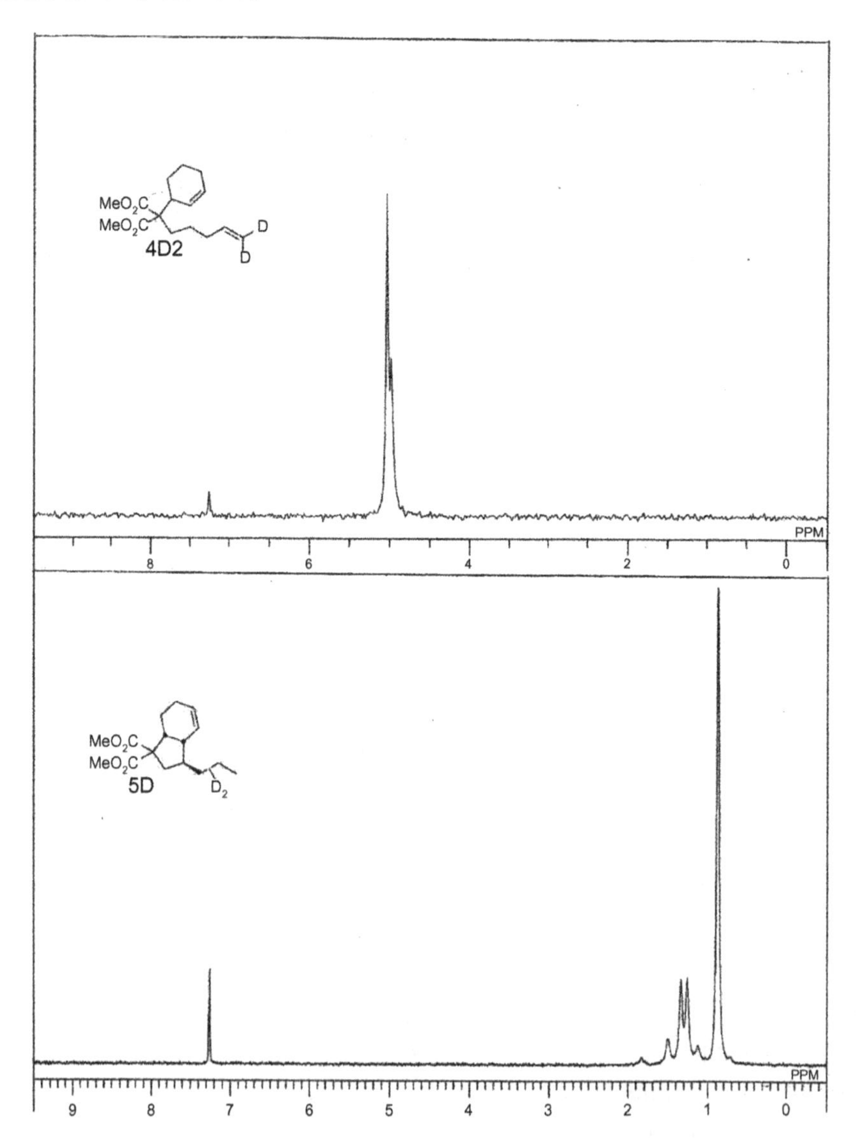

Figure 4.4 ^2H NMR spectra in $CHCl_3$ of deuterium-labeled diene and cyclization product. Source: Reprinted from: Hamasaki T, Aoyama Y, Kawasaki J, Kakiuchi F, Kochi T. *J Am Chem Soc* 2015;**137**:16163. Copyright 2015 American Chemical Society.

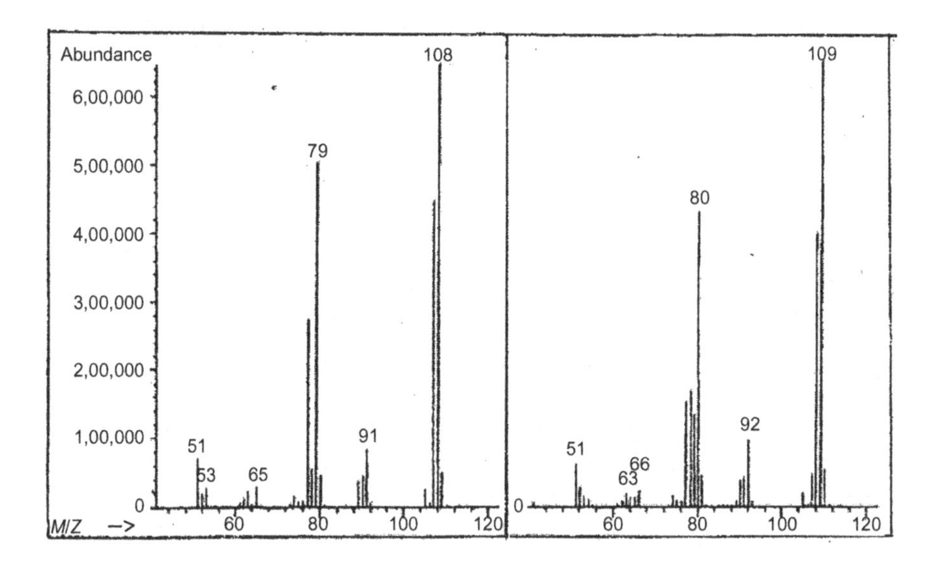

Figure 4.5 GC/MS spectra of ordinary and monodeuterated benzylalcohols. Source: Reprinted from: Robinson RK, De Jesus K. *J Chem Ed* 1996;**73**:264. Copyright 1996 American Chemical Society.

The mass spectra of the two benzylalcohols clearly establish the source of hydrogens in the reduction products (Fig. 4.5).

It is to be noted that the molecular ion peak ($M^+ = 109$) of deuterated benzyl alcohol is an odd number. This is a special case of deuterium-labeled compounds. When the compound contains an odd number of deuterium, the conventional "nitrogen rule" does not apply.

Organic chemists have been using deuterium beginning as early as in 1939. The applications range from KIEs to a tracer study to total synthesis of natural products. Selected examples from the literature will be presented next to illustrate the usefulness of deuterium and the deep impact of Urey's discovery of deuterium on organic chemistry.

4.2 CLASSIC EXAMPLES

The deuterium isotope effect has been found to be of great value in the study of chemical reactions and in the development of the theories of rate processes.

4.2.1 Reaction #1: Bromination of Acetone

In 1939, Reitz and Kopp reported their study on the acid-catalyzed bromination of acetone (Scheme 4.7).[12] An outstanding KIE was observed: ordinary acetone reacted seven times faster than deuterated acetone-d_6.

$$H_3C{-}CO{-}CH_3 + Br_2 \xrightarrow[25^\circ C,\ k_H]{H_3O^+,\ H_2O} H_3C{-}CO{-}CH_2Br$$

$$D_3C{-}CO{-}CD_3\ (92\%\ D) + Br_2 \xrightarrow[25^\circ C,\ k_D]{H_3O^+,\ H_2O} D_3C{-}CO{-}CD_2Br$$

$$\frac{k_H}{k_D} = 7.7$$

Scheme 4.7 Acid-catalyzed bromination.

The kinetic data suggest that the C–H bond breaking to form the corresponding enol is the rate-determining step of the reaction. The overall mechanism is therefore as follows (Scheme 4.8).[13]

$$H_3C{-}CO{-}CH_3 + H^+ \overset{fast}{\rightleftharpoons} H_3C{-}C(=\overset{+}{O}H){-}CH_3 \quad (1)$$

$$H{-}CH_2{-}C(=\overset{+}{O}H){-}CH_3 \overset{slow}{\rightleftharpoons} CH_2{=}C(OH){-}CH_3 + H^+ \quad (2)$$

$$CH_2{=}C(OH){-}CH_3 + Br_2 \overset{fast}{\rightleftharpoons} Br{-}CH_2{-}C(=\overset{+}{O}H){-}CH_3 + Br^- \quad (3)$$

$$Br{-}CH_2{-}C(=\overset{+}{O}H){-}CH_3 \overset{fast}{\rightleftharpoons} Br{-}CH_2{-}CO{-}CH_3 + H^+ \quad (4)$$

Scheme 4.8 Mechanism.

4.2.2 Reaction #2: Jones Oxidation

Jones oxidation of primary and secondary alcohols with chromium (VI) reagent gives aldehydes and ketones, respectively in high yields (Scheme 4.9).[14]

$$CH_3CH_2CH_2OH \xrightarrow[H_2O\ (90\%)]{H_2CrO_4} CH_3CH_2CHO$$

$$\text{cyclooctanol} \xrightarrow[H_2O,\ acetone\ (95\%)]{H_2CrO_4} \text{cyclooctanone}$$

Scheme 4.9 Jones oxidation.

To understand the reaction mechanism, Westheimer and Nicolaides performed a kinetic study of chromic acid oxidation of isopropyl alcohol to acetone using deuterium-labeled 2-propanol (Scheme 4.10).[15] The requisite 2-deuterio-2-propanol was prepared by the catalytic deuteration using deuterium gas in the presence of Adams catalyst promoted with a few crystals of ferrous chloride. Although the 2-D-2-propanol had relatively low deuterium content, it was good enough to be used in the kinetic study.

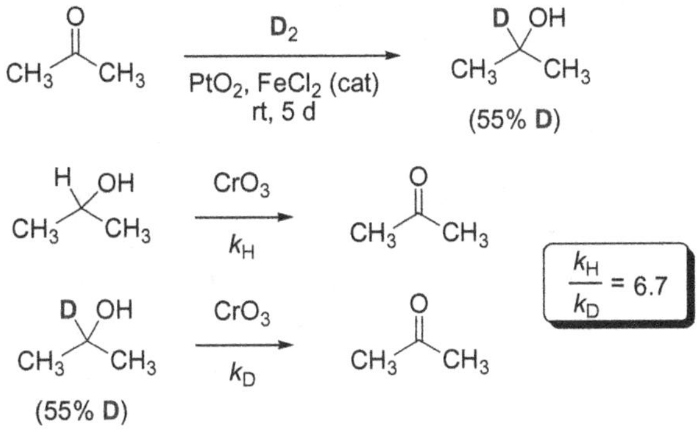

Scheme 4.10 Kinetic isotope effect.

The oxidation rates for 2-propanol and 2-D-2-propanol (55% D at C-2) were determined at 40°C by following the quantity of chromic acid remaining in solution using spectrophotometer. The ratio of the oxidation rates for 2-propanol and 2-D-propanol (55% D) was calculated to be 6.7 after considering the fact that the deuterium content of the 2-D-2-propanol was only 55%.

The kinetic results prove that the removal of the hydrogen is the rate-determining step in the Jones oxidation reaction (Scheme 4.11).

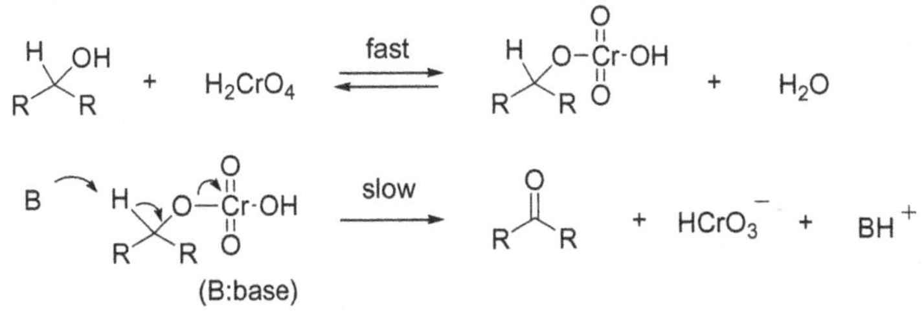

Scheme 4.11 Mechanism.

4.2.3 Reaction #3: E2 Elimination

Treatment of alkyl halide (RX) by a strong base such as potassium ethoxide produces an alkene via an elimination of hydrogen halide (HX). The reaction, known as bimolecular elimination, or E2 reaction, is usually considered a completely synchronous process with no intermediates detected (Scheme 4.12).[16]

$CH_3CH_2CBr(CH_3)_2$ —KOEt, EtOH→ $(CH_3)_2C{=}CHCH_3$ + $CH_3CH_2C(CH_3){=}CH_2$ (71:29)

Scheme 4.12

In 1952, Shiner examined the validity of the E2 reaction mechanism by determining the isotope effect on the reaction rate when the β-hydrogen atoms of isopropyl bromide were replaced by deuterium (Scheme 4.13).[17] The isopropyl bromide-d_6 was prepared in two steps: reduction of acetone-d_6 with lithium aluminum hydride gave isopropyl alcohol-d_6, which underwent reaction with phosphorus bromide to afford isopropyl bromide-d_6 of high D-content.

$D_3C(CO)CD_3$ —$LiAlH_4$, Diethyl carbitol (75%)→ $D_3CCH(OH)CD_3$ —PBr_3 (60%)→ $D_3CCHBrCD_3$ (98.5% D)

Scheme 4.13 Deuterium-labeled substrate.

If a reaction involves the removal of a hydrogen atom in its rate-determining step, it will show a marked difference in rate for removal of protium or deuterium. The kinetic studies were performed by treating ordinary or deuterium-labeled isopropyl bromides with sodium ethoxide in ethanol (Scheme 4.14).

$H_3CCHBrCH_3$ —NaOEt, EtOH, 25°C, k_H→ $H_3CCH{=}CH_2$

$D_3CCHBrCD_3$ —NaOEt, EtOH, 25°C, k_D→ $D_3CCH{=}CD_2$

$$\frac{k_H}{k_D} = 6.7$$

Scheme 4.14 Kinetic isotope effect in E2 reaction.

The kinetic data showed that the elimination reaction slowed down considerably for isopropyl bromide-d_6 compared to ordinary isopropyl bromide. It was concluded that the β-hydrogen elimination is the rate-determining step of the reaction.

One quick note should be made on the substitution reaction. When isopropyl bromide was treated with sodium ethoxide, an S_N2 reaction product was formed as a minor product (Scheme 4.15). No KIE was observed, indicating that the αC–H bond is not involved in the transition state.[17]

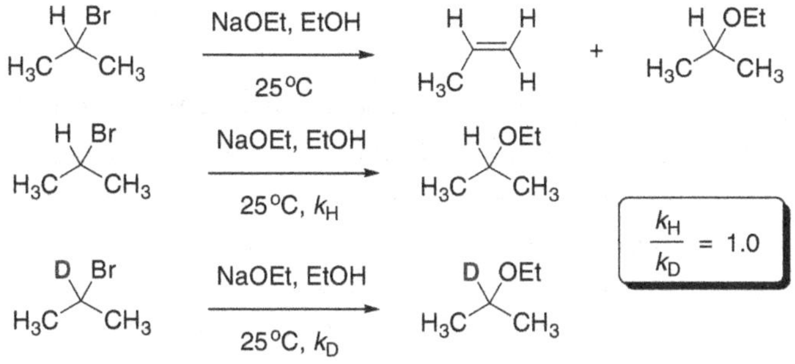

Scheme 4.15 Kinetic isotope effect in S_N2 reaction.

Another example of E2 reaction is a study carried out by Saunders and Edison (Scheme 4.16).[18] Decarboxylation of the sodium salt of phenylmalonic ester in deuterium oxide followed by $LiAlH_4$ reduction gave 2-phenylethanol-2,2-d_2. A sequence of O-tosylation and bromide substitution reaction afforded the required bromide **6D**.

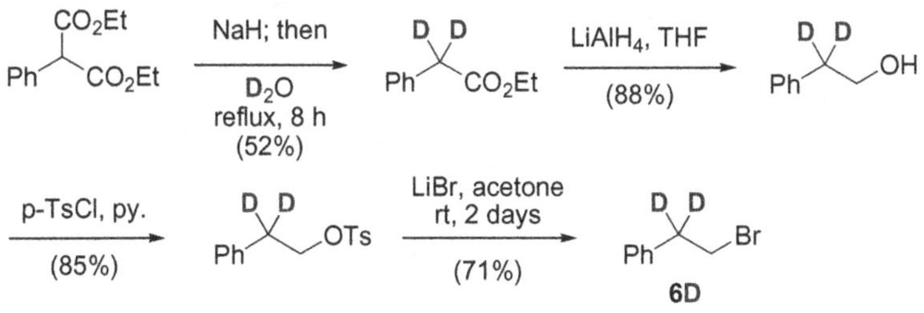

Scheme 4.16 Deuterium-labeled substrate.

With deuterium-labeled substrate in hand, E2 elimination reaction was performed in ethanol with sodium ethoxide as a base (Scheme 4.17). As expected, a large KIE of 7.11 was observed. The results confirmed that the proton abstraction is the rate-determining step in the E2 reaction.

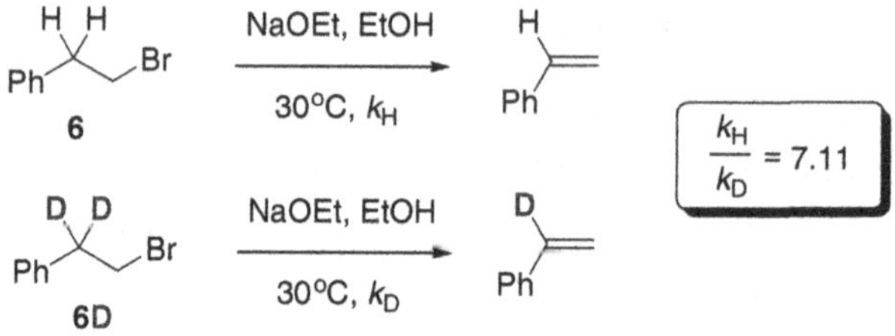

Scheme 4.17 Kinetic isotope effect.

4.2.4 Reaction #4: Reduction of Alkyl Halides by $LiEt_3BH$

Lithium triethylborohydride ($LiEt_3BH$, Super-Hydride), which can be prepared from lithium hydride and triethylborane, is a powerful nucleophile in displacement reactions of organic halides (Scheme 4.18).[19] Kinetic studies reveal that the reagent is 20 times more nucleophilic than thiophenoxide. The power of the reagent was clearly demonstrated by reduction of neopentyl bromide: the usually sluggish bromide underwent reaction in 3 h at reflux to afford the product in 96% yield!

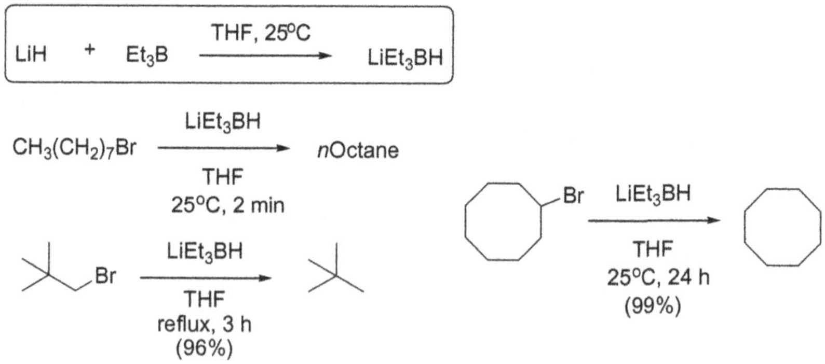

Scheme 4.18

To prove that the reaction is a nucleophilic substitution reaction, a reaction was carried out on exo-2-bromonorbornane using a deuterium-labeled reagent, $LiEt_3BD$ (Scheme 4.19). Lithium triethylborodeuteride was readily prepared from lithium deuteride and triethylborane in THF.

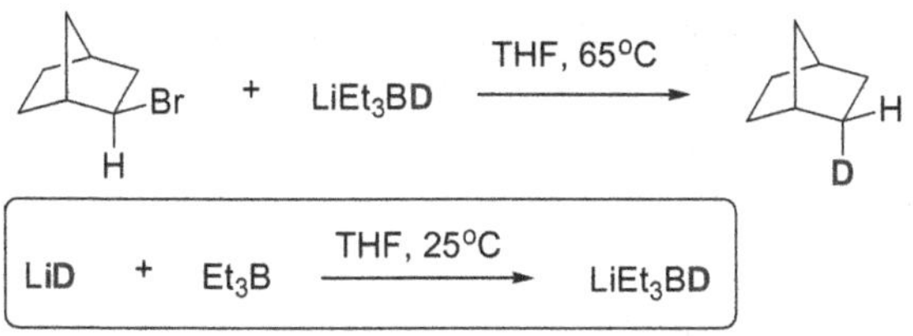

Scheme 4.19 Reaction with $LiEt_3BD$.

The product was identified as the pure endo isomer by inspecting the ^1H NMR spectrum of the product: while the relative intensity of exo protons (δ 1.43 ppm) to the bridgehead protons remained unchanged (4:2), the intensity of the signal due to the endo protons (δ 1.13 ppm) diminished (3:2). This proved that the reaction was an S_N2 resulting in a clean stereochemical inversion at the substitution center.

4.2.5 Reaction #5: Hydroboration of Alkenes

The hydroboration of alkenes is an important reaction in organic chemistry due to the synthetic versatility of the resultant organoborane intermediates, which can be transformed into a variety of useful functional groups. For example, hydroboration of 1-methylcyclopentene, followed by oxidation–hydrolysis work-up led to a high yield formation of *trans*-2-methylcyclopentanolmethylcyclopentanol (Scheme 4.20, Eq. 1), while hydroboration of styrene, followed by protonolysis, produced ethylbenzene in 88% yield (Scheme 4.20, Eq. 2).[20]

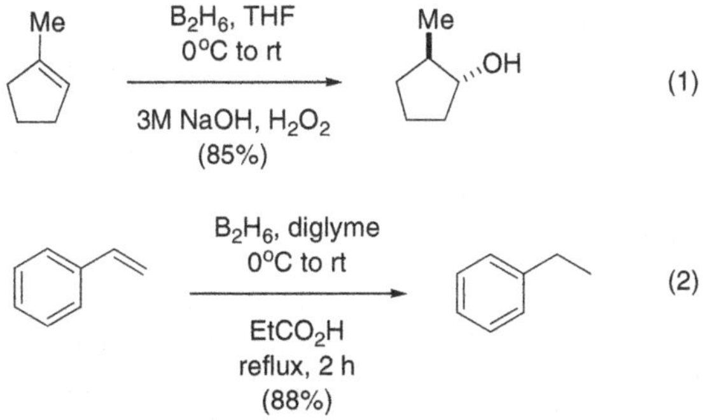

Scheme 4.20

Brown and coworkers carried out an extensive study to provide an evidence for the retention of configuration in the protonolysis of organoboranes, which formed by reaction of diborane with alkenes. In doing so, a series of deuterium-labeled reagents were employed (Scheme 4.21).[21] The hydroboration of norbornene readily produced the corresponding organoborane intermediate **A**, which upon protonolysis with propionic acid formed norbornane.

B_2H_6 / $EtCO_2H$ → **7** (H^e, H^a) via [**A** (H^e, BR_2)]

B_2H_6 / $EtCO_2D$ → **7a** (H, D)

B_2D_6 / $EtCO_2H$ → **7b** (D, H)

B_2D_6 / $EtCO_2D$ → **7c** (D, D)

Scheme 4.21 Deuterium-labeling experiments.

The ^{1}H NMR spectrum of norbornane indicated four equatorial hydrogen atoms. The monodeuterated norbornanes (**7a** and **7b**) exhibited the same ^{1}H NMR spectra as norbornane's, except that the number of equatorial hydrogen atoms decreased to 3. Finally, the NMR spectrum of dideuterated derivative **7c** obtained from a deuterioboration/deuterolysis sequence showed only two equatorial hydrogen atoms. These results are consistent both with the exclusive *syn* addition of the boron–hydrogen (boron–deuterium) bond from the less hindered *exo* face and with the retention of configuration in the protonolysis (deuterolysis) of the organoborane intermediate **A**.

Herbert C. Brown received the 1979 Nobel Prize in chemistry for his development of the use of boron-containing compounds into important reagents in organic synthesis[22]:

> *Brown was born on May 22, 1912 in London, United Kingdom. In 1914, his father decided to move to Chicago to join his family members. He entered the University of Chicago in the Fall of 1935 and graduated with BS degree in 1936. Upon graduation, he received a copy of Alfred Stock's book,* The Hydrides of Boron and Silicon, *which interested him in the chemistry of*

boron to begin graduate study with Professor H. I. Schlesinger. He obtained his PhD degree in 1938. After a brief postdoctoral stint with Professor Morris S. Kharasch, he became a research assistant for Professor Schlesinger. He began his independent academic career at Wayne State University in Detroit. He became associate professor in 1946 and was invited to Purdue University in 1947. He became Wetherill Distinguished Professor in 1959. His major research program was in the borane–organoborane area. He was elected to the National Academy of Sciences in 1957, the American Academy of Arts and Sciences in 1966. He received the ACS Award for Creative Research in synthetic organic chemistry in 1960. He died on December 19, 2004 in Lafayette, IN.

Although the reaction is believed to involve the *syn* addition of the boron and hydrogen moieties to an alkene, Kabalka and coworkers wanted to confirm the stereochemistry of hydroboration using deuterium-labeled substrates (Scheme 4.22).[23] Deuterioboration of 1-hexyne with di(2-deuteriocyclohexyl)borane-B-d_1 gave (*E*)-1-hexene-1,2-d_2 **8a***E* after deuterolysis with acetic acid-d_1. On the other hand, hydroboration of 1-hexyne with dicyclohexylborane followed by deuterolysis afforded (*E*)-1-hexene-d_1 **8b***E*. Two other alkenes, **8a***Z* and **8b***Z*, were prepared in a similar manner starting from 1-hexyne-1-d_1.

$(c\text{-}C_6H_{10}D)_2BD$, CH_3CO_2D: *n*Bu–C≡C–H → **8a***E*; $(c\text{-}C_6H_{11})_2BH$, CH_3CO_2D: *n*Bu–C≡C–H → **8b***E*

$(c\text{-}C_6H_{10}D)_2BD$, CH_3CO_2H: *n*Bu–C≡C–D → **8a***Z*; $(c\text{-}C_6H_{11})_2BH$, CH_3CO_2H: *n*Bu–C≡C–D → **8b***Z*

Scheme 4.22 Deuterium-labeled substrates.

Both alkenes, **8a***E* and **8a***Z*, were treated with ordinary dicyclohexylborane, and the other two alkenes, **8b***E* and **8b***Z*, were reacted with di(2-deuteriocyclohexyl)borane-B-d_1 (Scheme 4.23). If addition of dialkylborane was to occur in a *syn* mode, *threo*-isomer **9t** would form from both **8a***E* and **8b***Z* and *erythro*-isomer **9e** would form from both **8a***Z* and **8b***E*. ^{1}H NMR analysis of the hydroboration products confirmed that this was the case. The study therefore provided direct evidence that the hydroboration reaction involves the *syn* addition of the boron and hydrogen moieties to the alkene.

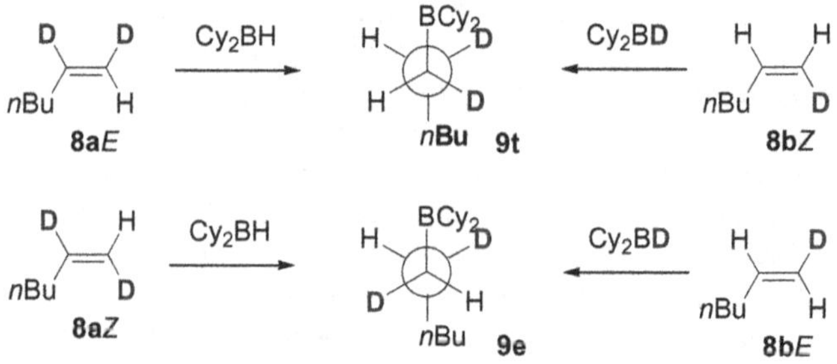

Scheme 4.23 Reaction with deuterium-labeled substrates.

4.2.6 Reaction #6: Reaction of *cis*-Cyclodecene Oxide with Lithium Diethylamide

In 1960, Cope and coworkers reported that reaction of *cis*-cyclodecene oxide **10** with lithium diethylamide in hot benzene produced *cis-cis*-1-decalol **11a** as a major product along with two minor products (**11b** and **11c**) in 77% combined yield (Scheme 4.24).[24]

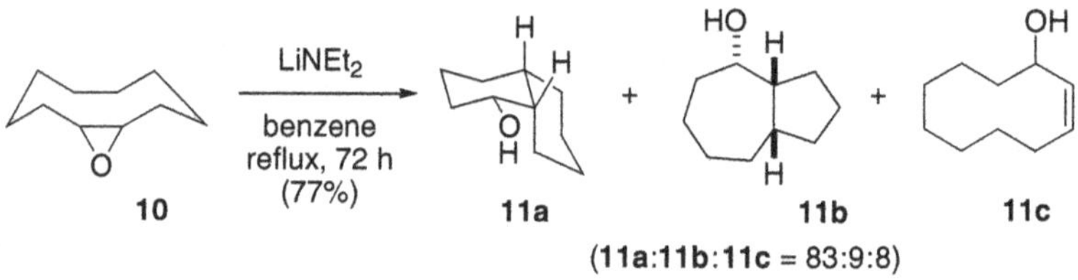

Scheme 4.24

Two reaction pathways leading to the bicyclic alcohol **11a** appeared to be possible (Scheme 4.25). In path *a*, the base could remove a proton from a carbon located across the ring from the epoxide ring with concerted opening of the epoxide by the carbanion so formed. In path *b*, the base removes a proton from one of the carbon atoms of the epoxide ring, followed by a breaking of the bond between the oxygen atom and the carbon atom which has lost a proton to form a carbene intermediate, which could then form a bond to a carbon atom across the ring with simultaneous transfer of hydride ion from the same carbon atom to the electron-deficient carbon atom.

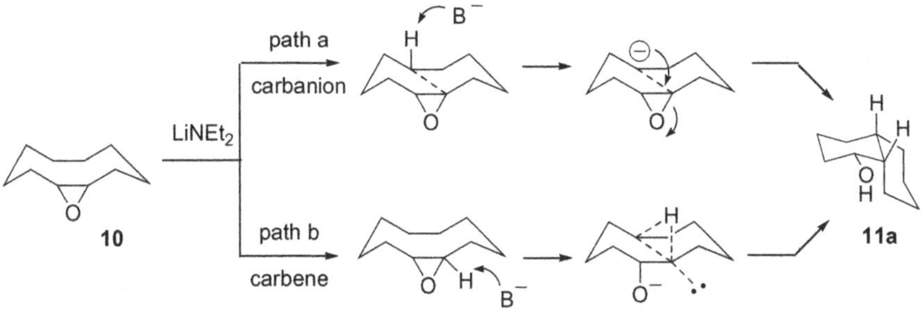

Scheme 4.25 Two possible mechanisms.

To determine which pathway is at work, deuterium tracer studies were carried out using the *cis*-cyclodecene oxide-d_2 **10D** (Scheme 4.26). Deuteroboration of cyclodecyne with diborane-d_6 followed by quenching with acetic acid-d exclusively gave the desired *cis*-isomer, which underwent epoxidation with monoperphthalic acid to produce the epoxide **10D** in 85% yield and 89% deuterium enrichment.

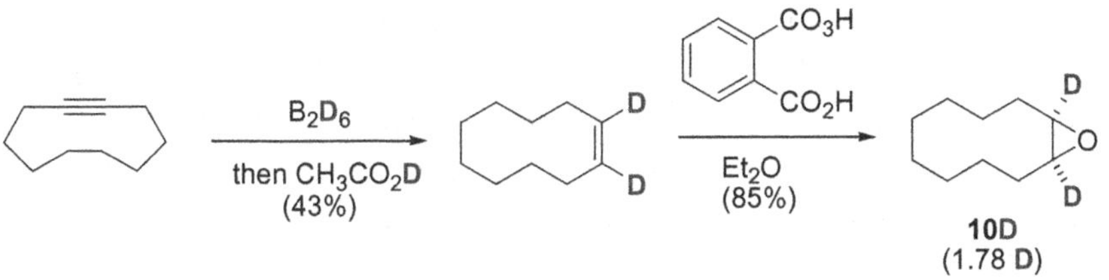

*Scheme 4.26 Synthesis of epoxide **10D**.*

When the epoxide **10D** was treated with lithium diethylamide under usual conditions, it was found that the *cis-cis*-1-decalol contained 0.96 atom of deuterium per molecule (Scheme 4.27). Thus path *b* appears to be the route by which the bicyclic alcohol is formed. The correct structure of *cis-cis*-1-decalol is then **11aD1**.

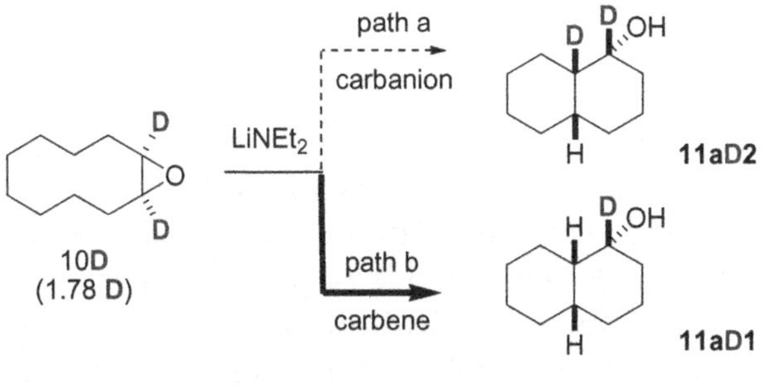

Scheme 4.27 Deuterium-labeling experiments.

Arthur Cope was professor of chemistry at Massachusetts Institute of Technology[25]:

Arthur Clay Cope was born on June 27, 1909, in Dunreith, IN. In 1929, he obtained the bachelor's degree in chemistry from Butler University in Indianapolis. He studied under S. M. McElvain at the University of Wisconsin and graduated in 1932. A year later, he moved to Harvard University to work under E. P. Kohler. In 1934, he began his academic career at Bryn Mawr College, where he developed the facile thermal rearrangement, now known as "the Cope rearrangement." After a brief stay at Columbia University, he came to MIT to head the Department of Chemistry in 1945. The same year he was appointed to the editorial board of Organic Syntheses, *which Roger Adams had founded. He was elected to the National Academy of Sciences in 1947 and President of the American Chemical Society in 1960. He died in Washington, DC, on June 4, 1966. He left half of his estate to the American Chemical Society, which administers the Cope Awards annually to recognize and encourage excellence in organic chemistry.*

4.2.7 Reaction #7: Hofmann–Löffler Reaction

The Hofmann–Löffler reaction is a conversion of *N*-bromo or *N*-chloroamines to the cyclic amines.[26] For example, Löffler and Kober reported an elegant synthesis of nicotine starting from the amine **12** in 1909 (Scheme 4.28, Eq. 1).[27] L-Proline was synthesized by the same reaction starting from *N*-chloro-L-norvaline **13** (Scheme 4.28, Eq. 2).[28]

Br N Me 12 — 1. H_2SO_4, 140°C; 2. OH^- → N Me Nicotine (1)

Cl N H Me CO_2Et 13 — 1. H_2SO_4, 140°C; 2. OH^- → CO_2H N H L-proline (2)

Scheme 4.28

Because the formation of pyrrolidines is mechanistically a very unusual transformation, Corey and coworker decided to investigate the reaction using deuterium-labeled substrates (Scheme 4.29).[29] $LiAlD_4$ reduction of the sulfonamide **14** gave (−)-methylamylamine-4-d **15D** in 73% yield after acidic hydrolysis. Chlorination of the amine gave the corresponding *N*-chloroamine **16D**, which underwent the

Hofmann–Löffler reaction to afford 1,2-dimethypyrrolidine **17/17D** in 43% yield.

CH_3 TsO H N CH_3 Ts **14** — LiAlD$_4$, THF; 48% HBr, PhOH (73%, 2 steps) → CH_3 H D N CH_3 H **15D** — Cl_2 →

CH_3 H D N CH_3 Cl **16D** — H_2SO_4, 95°C (43%) → **17** + **17D** (**17**:**17D** ~ 1:3.54)

Scheme 4.29 Deuterium-labeling experiments.

Combustion analysis of the mixture of deuterated and nondeuterated 1,2-dimethylpyrrolidines showed the presence of 0.78 deuterium per molecule, which corresponds to an isotope effect of $k_H/k_D = 3.54$. Another important fact that came out of the experiment was that the pyrrolidine was optically inactive. This is strong evidence that the decomposition of *N*-chloroamine **16D** in acid involves an intermediate, in which $C\delta$ is trigonal.

Elias James Corey received the 1990 Nobel Prize in chemistry for his development for the theory and methodology of organic synthesis[30]:

Professor Corey was born on July 12, 1928, in Methuen, MA. He lost his father when he was just 18 months old. He entered the Massachusetts Institute of Technology in July 1945 when he was 16. At MIT, he found organic chemistry fascinating and had many superb teachers including Arthur C. Cope and John C. Sheehan. He obtained his PhD degree in 1950 at the age of 22, then joined the University of Illinois at Urbana-Champaign as an Instructor in Chemistry under the distinguished chemists Roger Adams and Carl S. Marvel. By 1954, he was able to initiate more complex projects dealing with stereochemistry and focusing on the synthesis of natural products. He became professor of chemistry at the age 27. In the spring of 1959, he moved to Harvard University. In the mid-1960s, he completed first chemical synthesis of prostaglandins using the concept of retrosynthetic analysis. Other outstanding examples are total syntheses of gibberellic acid and (+)-ginkolid. His research areas include synthesis of complex molecules, the logic of chemical synthesis, development of new methods of synthesis, theoretical organic chemistry and reaction mechanisms, and organometallic chemistry. He is currently Sheldon Emory Professor Emeritus. In September 1961, he married Claire Higham and they have three children.

4.2.8 Reaction #8: Diels–Alder Reaction

Diels–Alder reaction is the cycloaddition of 1,3-dienes and alkenes to form substituted cyclohexenes. For example, a thermal reaction of 1,3-butadiene with maleic anhydride in benzene gave the bicyclic compound (Scheme 4.30).[31]

benzene
100°C

Scheme 4.30

To understand the exact mechanism of the Diels–Alder reaction of 1,3-butadiene with ethylene, Houk and coworkers prepared deuterium-labeled substrates (Scheme 4.31).[32] Repeated H/D exchange reaction of sulfolene with D_2O in *p*-dioxane in the presence of K_2CO_3 afforded sulfolene-d_4, which upon heating gave 1,1,4,4,-tetradeuterio-1,3-butadiene **18D**. Stereospecific reduction of acetylene-d_2 afforded *cis*- and *trans*-ethylene: copper-activated zinc treatment in acidic water gave *cis*-ethylene-d_2 **19c**, whereas chromous chloride reduction provided pure *trans*-ethylene-d_2 **19t**.

K_2CO_3
D_2O, *p*-dioxane
rt, 48 h
repeat 5 times
130–162°C, 1 h
(88%)
18D
Zn–Cu
aq. HCl
$CrCl_2$
aq. HCl
19c
19t

Scheme 4.31 Deuterium-labeled substrates.

With deuterium-labeled substrates in hand, a Diels–Alder reaction was run at 185°C for 36 h at a pressure of 1800 psi in a stainless steel bomb (Scheme 4.32). Each cyclohexene product was separated and then subjected to epoxidation with *m*-chloroperoxybenzoic acid (*m*-CPBA) to determine the structures.

Scheme 4.32 Reaction with deuterium-labeled substrates.

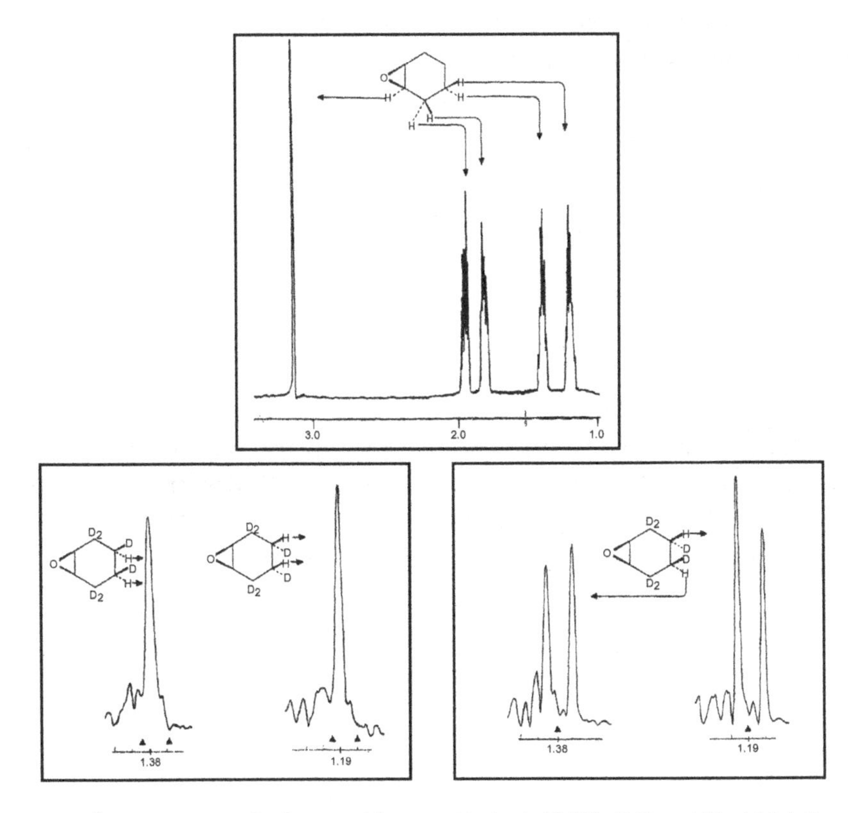

*Figure 4.6 ¹H NMR spectra of ordinary cyclohexene oxide (top), **20a/20b** (left), and **20c** (right).* Source: Reprinted from: Houk KN, Lin Y-T, Brown FK. *J Am Chem Soc* 1986;**108**:554. Copyright 1986 American Chemical Society.

Analysis of ^{1}H NMR spectra showed that there is less than 1 percent of the *trans*-adduct in the *cis*-products (**20a/20b**) and vice versa (Fig. 4.6).

If a diradical intermediate such as **A** was formed, an extensive scrambling of stereochemistry would occur. Thus the deuterium-labeling experiments demonstrated that Diels–Alder reaction of butadiene with ethylene occurs through a synchronous concerted mechanism.

4.2.9 Reaction #9: Tishchenko Reduction

Treatment of β-hydroxy ketone **21** with 4 equivalents of acetaldehyde or benzaldehyde and 15 mol% SmI_2 in THF at −10°C resulted in a rapid formation of *anti*-1,3-diol monoesters **22a** or **22b** in great yields and with superb diastereoselectivity (Scheme 4.33).[33] The diastereoselectivity observed in the reaction was explained by a six-membered transition state (**TS A**), in which samarium coordinates to both carbonyl and hemiacetal oxygens.[34]

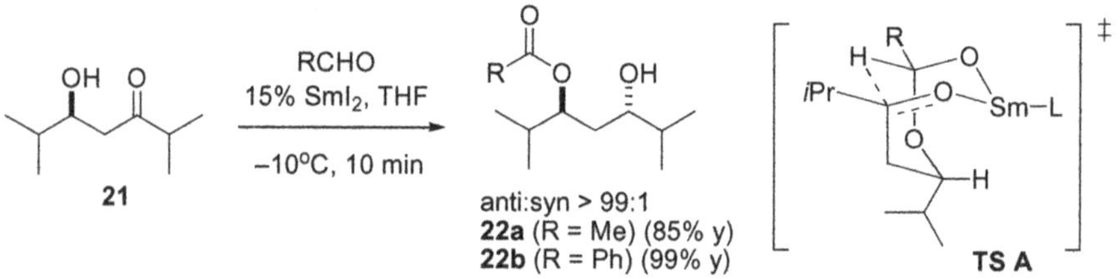

Scheme 4.33

To determine the source of hydride, a reaction was run with acetaldehyde-d_4 (Scheme 4.34, Eq. 1). Complete deuterium incorporation was observed at the newly generated carbinol center, confirming that the aldehyde is the exclusive source of hydride.

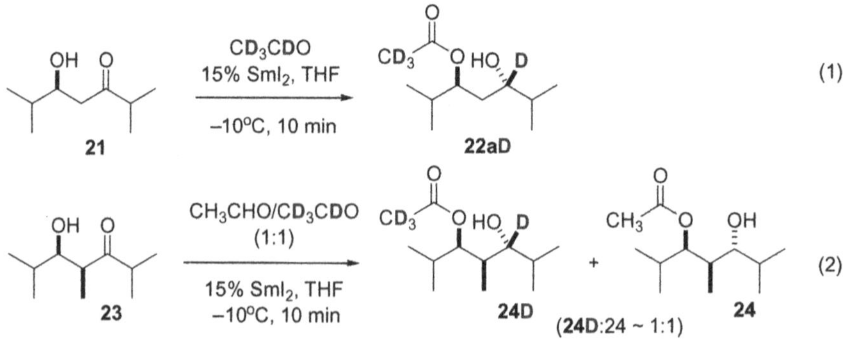

Scheme 4.34 Deuterium-labeling experiments.

A competition reduction of α-methyl-β-hydroxy ketone **23** using an equimolar mixture of CH_3CHO and CD_3CDO resulted in the formation of a 1:1 mixture of the deuterated and nondeuterated reduction products, implying that hydride transfer is not the rate-determining step (Scheme 4.34, Eq. 2).

4.2.10 Reaction #10: Asymmetric Isomerization of Allylamines to Enamines

The cationic rhodium complex of 2,2'-bis(diphenylphosphino)-1,1'-binaphthyl (BINAP) catalyzes asymmetric isomerization of diethylgeranylamine **25** to give citronellal (*E*)-ethyleneamine **26** in greater than 95% ee. This highly enantioselective reaction is a key step in the production of (−)-menthol (Scheme 4.35).[35]

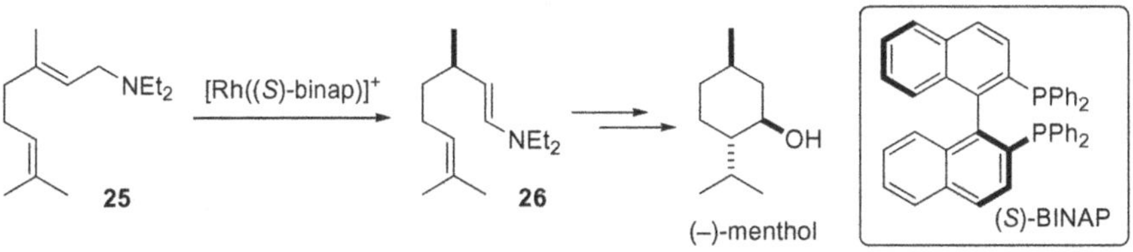

NEt2 [Rh((*S*)-binap)]+ 25 NEt2 26 OH (−)-menthol PPh2 PPh2 (*S*)-BINAP

Scheme 4.35

To understand the mechanism of the reaction, Noyori and coworkers prepared a deuterium-labeled substrate **25D** (Scheme 4.36). $LiAlD_4$ reduction of ethyl geranate produced geraniol-d_2 in quantitative yield. Oxidation followed by asymmetric reduction afforded geraniol-d **27**. Mitsunobu amination, deprotection, reductive alkylation gave the substrate **25D** in 93% ee and 98.6% deuterium content.

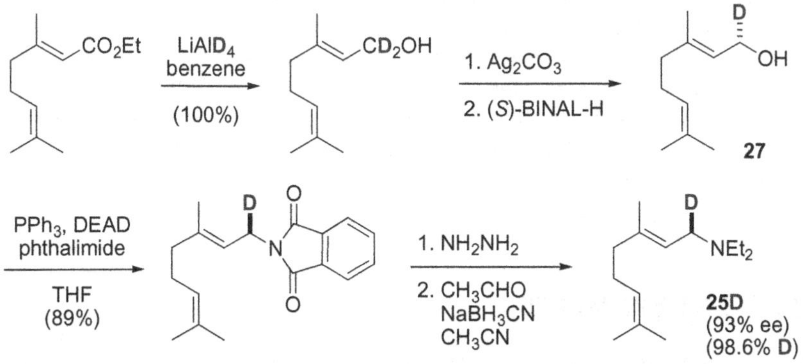

CO2Et LiAlD4 benzene (100%) CD2OH 1. Ag2CO3 2. (*S*)-BINAL-H D OH 27

PPh3, DEAD phthalimide THF (89%) D O N O 1. NH2NH2 2. CH3CHO NaBH3CN CH3CN D NEt2 25D (93% ee) (98.6% D)

Scheme 4.36 Deuterium-labeled substrate.

When geranylamine-d **25D** was treated with a catalytic amount of chiral catalyst from (*S*)-BINAP, a clean migration of C-1 protium took place to give (*R*,*E*)-enamine **26Da** (Scheme 4.37). Moreover, the same treatment with the chiral catalyst from (*R*)-BINAP, produced an isotopomer **26Db**. The 1,3-protium (deuterium) shift is occurring in a suprafacial manner.

[Rh((*S*)-binap)]$^+$ THF 40°C, 24 h → **26Da**

25D

[Rh((*R*)-binap)]$^+$ THF 40°C, 24 h → **26Db**

Scheme 4.37 Deuterium-labeling experiment.

Thus the deuterium-labeling experiment established that the chiral rhodium catalyst could recognize one of the two enantiotopic hydrogens at C-1[36]:

> *Ryoji Noyori received the 2001 Nobel Prize in chemistry for the work on chirally catalyzed hydrogenation reactions. He shared the prize with William S. Knowles and K. Barry Sharpless. He was born in September 1938 in Kobe, Japan. He obtained his BS degree in 1961 from Kyoto University. In 1967, he received his doctorate degree. He carried out his postdoctoral research at Harvard University with Professor Corey in 1969. He began his independent career at Nagoya University in 1967. One of his early discoveries was the asymmetric catalysis in carbene reaction mediated by a chiral Schiff base–copper (II) complex. In the 1980s, he had success in asymmetric hydrogenation largely relying on the invention of BINAP chiral ligand. In 1995, he invented a range of ruthenium catalysts modified with a chiral β-amino alcohol that affect asymmetric transfer hydrogenation using 2-propanol as a hydrogen donor. He received numerous awards including The Chemical Society of Japan Award for Young Chemists in 1972, The Cope Award in 1997, and The Roger Adams Award in Organic Chemistry in 2001. He is the author of a book* "Asymmetric Catalysis in Organic Synthesis" *published in 1994 by John Wiley & Sons, Inc.*

4.3 RECENT EXAMPLES

Organic chemists have continued to use a variety of deuterium-labeled compounds in their research on reaction mechanisms and synthesis. In this section, some of the most recent examples are presented.

4.3.1 Reaction #11: Ruthenium-Catalyzed Coupling of Propargylic Ethers with Alcohols

In 2015, a team led by Professor Krische at the University of Texas, Austin, reported that a highly stereoselective coupling reaction of propargyl ether **28** with a variety of primary alcohols **29a–29c** could be achieved with the use of ruthenium catalyst to provide 1,4-diols **30a–30c** (Scheme 4.38).[37] Under the reaction conditions, the propargyl ether transformed to a crotyl metal reagent, and the primary alcohols became aldehydes.

28 (3 eq.)

29a, R = Ph
29b, R = Ph-4-Br
29c, R = *n*Hex

* Conditions:
Ru$(H_2)(CO)(PPh_3)$ (5 mol%)
$ArSO_3H$ (7.5 mol%)
Chiral ligand (5 mol%)
Bu_4NI, 2-PrOH (2 eq)
THF, 85°C, 24 h

Product	Yield, %	ee, %
30a	70	91
30b	77	92
30c	81	90

Scheme 4.38

This highly efficient reaction was used in a short synthesis of (+)-*trans*-whiskey lactone **32** (Scheme 4.39). Thus reaction of the propargyl ether **28** with 1-pentanol gave γ-lactol **31** after TBAF treatment of the crude product. PCC oxidation in the presence of sodium acetate gave the desired lactone in 73% overall yield and 90% ee.

Scheme 4.39 Synthetic application.

The researchers then turned their attention to the mechanism of the reaction, and decided to conduct deuterium-labeling studies. The required primary alcohols were prepared by lithium aluminum deuteride reduction of the esters (Scheme 4.40).

Scheme 4.40 Synthesis of deuterium-labeled substrate.

With deuterium-labeled substrates in hand, the reactions were repeated under standard conditions (Scheme 4.41). Reaction of deuterated propargyl ether **28D** with ordinary 4-bromobenzyl alcohol gave the coupling product **33Da**, in which complete deuterium enrichment at the C-1 and C-2 was observed. The results hint that a crotyl ruthenium species A is being generated in the reaction.

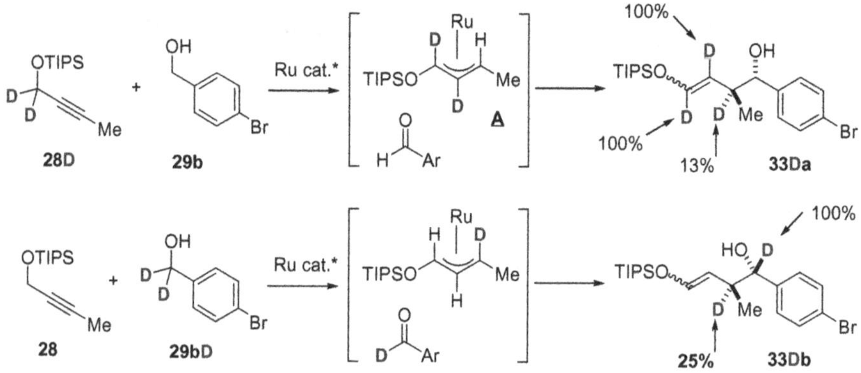

Scheme 4.41 Deuterium-labeling experiments.

Another reaction employing deuterated 4-bromobenzyl alcohol **29bD** delivered the product **33Db**, in which deuterium was found at C-4. The fact that no deuterium was present at the vinylic carbons excludes the possibility of allene hydrometallation.

Based on these results, a mechanism was proposed (Scheme 4.42). Coordination of ruthenium (0) to the alkyne gives the metallocyclopropene B, which upon 1,2-deuteride shift affords the vinyl carbene C. Protonation leads to an allylic ruthenium(II) complex A, which undergoes a nucleophilic addition to the aldehyde to deliver the product according to the Zimmerman–Traxler type six-membered transition state **TS A**.[38]

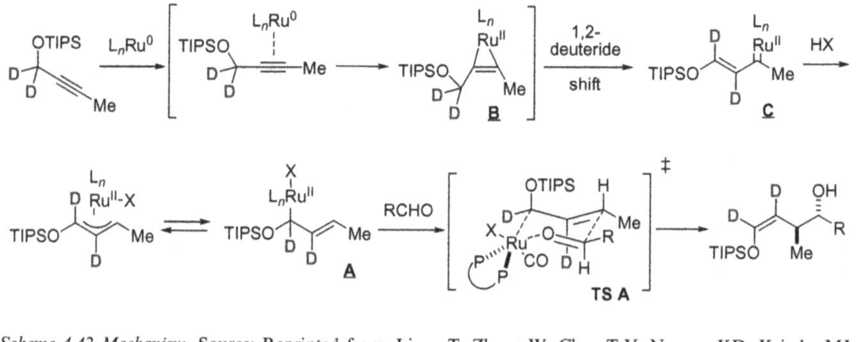

Scheme 4.42 Mechanism. Source: Reprinted from: Liang T, Zhang W, Chen T-Y, Nguyen KD, Krische MJ. *J Am Chem Soc* 2015;**137**:13066. Copyright 2015 American Chemical Society.

4.3.2 Reaction #12: Cobalt-Catalyzed (*Z*)-Selective Hydroboration of Terminal Alkynes

Professor Chirik and coworkers at Princeton University recently reported that a highly *Z*-selective hydroboration of terminal alkynes could be achieved in good yields with the use of a cobalt catalyst **A** (Scheme 4.43).[39]

Product	*Z*:*E*	Yield, %
35a	92:8	76
35b	97:3	91
35c	95:5	83

Scheme 4.43

As the *Z*-selectivity observed with the catalyst is uncommon, the researchers conducted a series of deuterium-labeling experiments. The requisite deuterium-labeled reagents were prepared (Scheme 4.44). Octyne-1-d **34aD** was prepared in good yield by heating a heterogeneous mixture of 1-octyne with 40% NaOD. Deuterated pinacolborane (DBPin) was obtained by the action of deuterium gas in the presence of a catalyst **B**.

Scheme 4.44 Synthesis of deuterium-labeled reagents.

The hydroboration was then repeated with deuterated reagents (Scheme 4.45). Hydroboration of deuterated 1-octyne with ordinary pinacolborane under standard reaction conditions gave (*Z*)-isomer

35DH*Z* as a major product with deuterium attached to the C-2 carbon. Another experiment employing ordinary 1-octyne with deuteriopinacolborane afforded almost exclusively (*Z*)-isomer **35HD***Z*, in which deuterium and pinacolborane are attached to the same carbon.

$CH_3(CH_2)_5C\equiv CD$ (**34aD**) + H–BPin (HBPin) → [cat. **A** (3 mol%), THF, 23°C, 6 h, 72% conv.] **35DH(*Z*)** : **35HD(*E*)** = 73:27 (1)

$CH_3(CH_2)_5C\equiv CH$ (**34a**) + D–BPin (DBPin) → [cat. **A** (3 mol%), THF, 23°C, 6 h, 98% conv.] **35HD(*Z*)** : **35DH(*E*)** = 93:7 (2)

Scheme 4.45 Reactions with deuterium-labeled reagents.

An interesting observation made for these two parallel reactions was that duterated 1-octyne underwent reaction slower than ordinary 1-octyne. The two regioisomers, **35HD***Z* and **35DH***E*, were readily distinguished by ^{2}H NMR (Fig. 4.7).

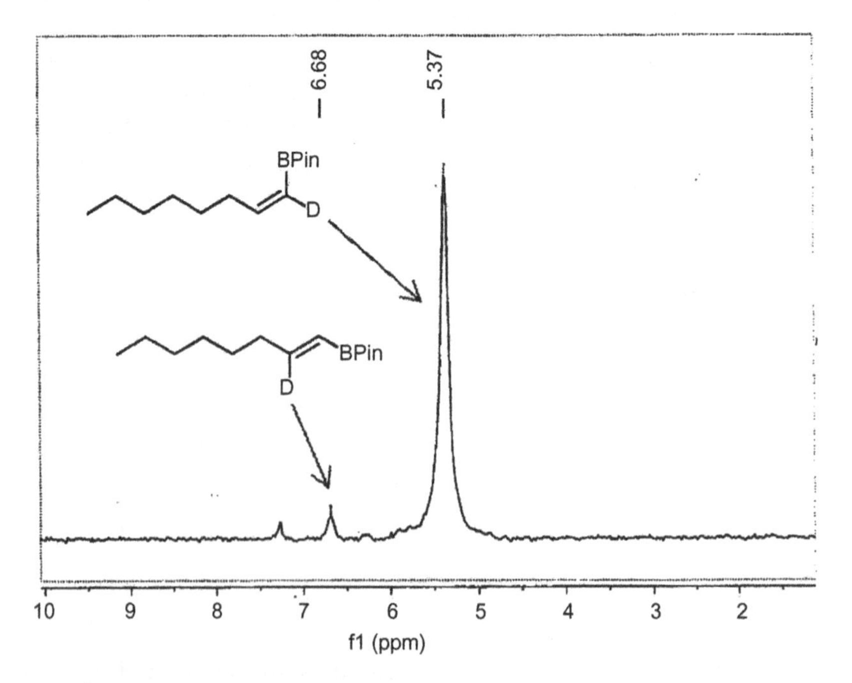

*Figure 4.7 ^{2}H NMR spectrum of a mixture of hydroboration products, **35HDZ** and **35DH**E.* Source: Reprinted from: Obligacion JV, Neely JM, Yazdani AN, Pappas I, Chirik PJ. *J Am Chem Soc* 2015;**137**:5855. Copyright 2015 American Chemical Society.

With the above results and other experimental data, a mechanism was proposed (Scheme 4.46). The cobalt catalyst **A** reacts with the alkyne to form the cobalt acetylide **B** with liberation of methane. Oxidative addition of DBPin gives **C**, which upon reductive elimination produces **D**. *syn*-Hydrocobaltation generates the (*Z*)-cobalt vinyl intermediate **E**, which delivers the (*Z*)-boronate ester upon reacting with the alkyne.

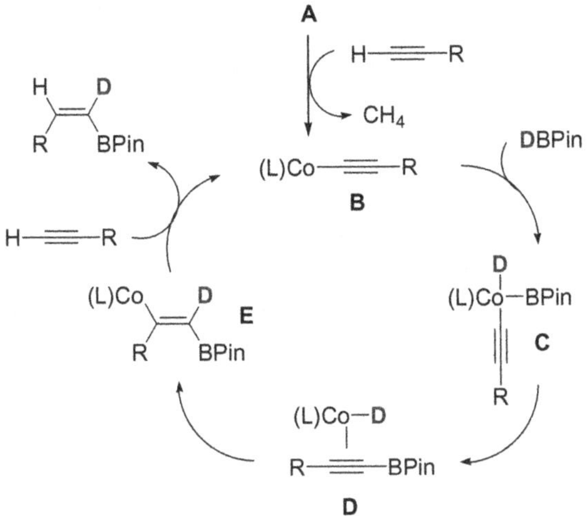

Scheme 4.46 Mechanism. Source: Reprinted from: Obligacion JV, Neely JM, Yazdani AN, Pappas I, Chirik PJ. *J Am Chem Soc* 2015;**137**:5855. Copyright 2015 American Chemical Society.

4.3.3 Reaction #13: Palladium-Catalyzed C–H Activation of Alkyl Arenes

John Curto and Professor Kozlowski of the University of Pennsylvania communicated in 2015 that phenylglycine azlactone **36** underwent a coupling reaction with a variety of arenes such as toluene and ethylbenzene to afford products (**37a**, **37b**) in good yields (Scheme 4.47).[40] The C–H activation reaction was chemoselective in that only the methyl C–H was responding to the palladium catalyst.

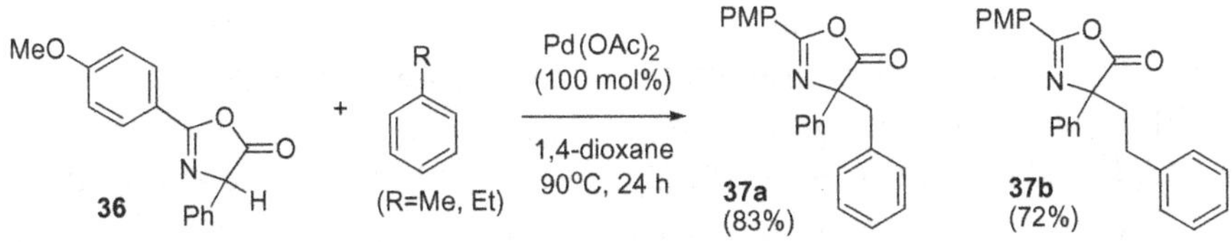

Scheme 4.47

To investigate the reaction mechanism, deuterium-labeling studies were conducted using an equimolar amount of ordinary toluene and toluene-d_8 (Scheme 4.48). A KIE of $k_H/k_D = 3.55$ was observed, indicating that the C–H activation is the rate-determining step.

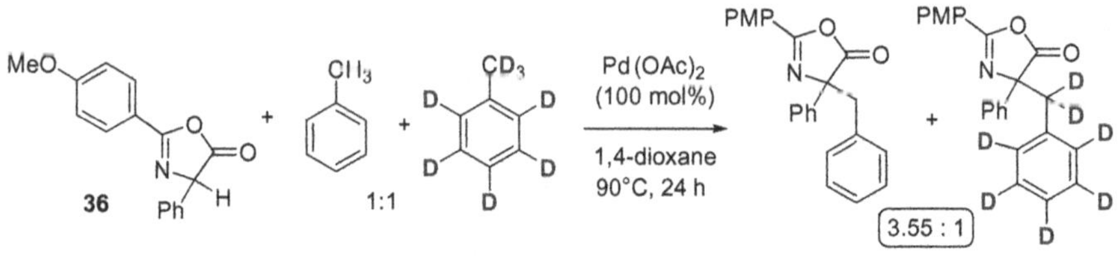

Scheme 4.48 Kinetic isotope effect.

Another deuterium-labeling study was run employing ethylbenzene-d_2 (Scheme 4.49). A deuterium scrambling occurred, which suggests an initial Pd metalation of the benzylic C–H (C–D) bond through intermediate **A**.

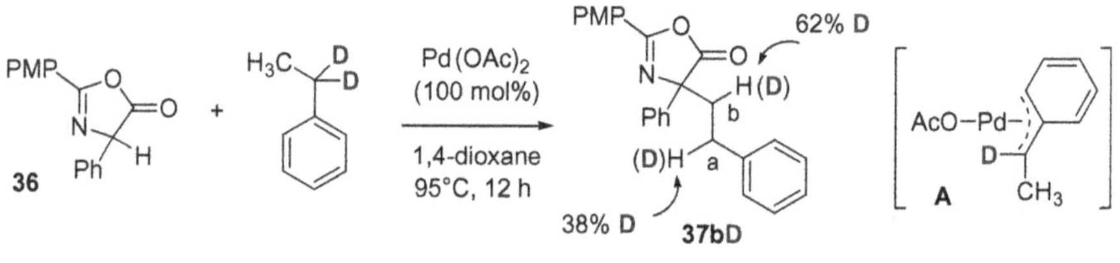

Scheme 4.49 Reactions with ethylbenzene-d_2.

4.3.4 Reaction #14: Reaction of Cyclohexanol with *o*-Benzynes

In studying reactions of secondary and primary alcohols with benzynes generated by the hexahydro-Diels–Alder reaction, Professor Hoye and coworkers at the University of Minnesota used deuterium-labeled substrates to decipher the reaction mechanism.[41] When the triyne **38** was heated in neat cyclohexanol **39HH**, the ether **40** was produced in 80% yield along with a small amount of the reduced benzenoid **41HH** in a 12:1 (**40**:**41HH**) ratio. Surprisingly the ratio dramatically reversed to 1:17 when a smaller amount (only 1.6 equivalents) of cyclohexanol was used in $CDCl_3$ solvent (Scheme 4.50).

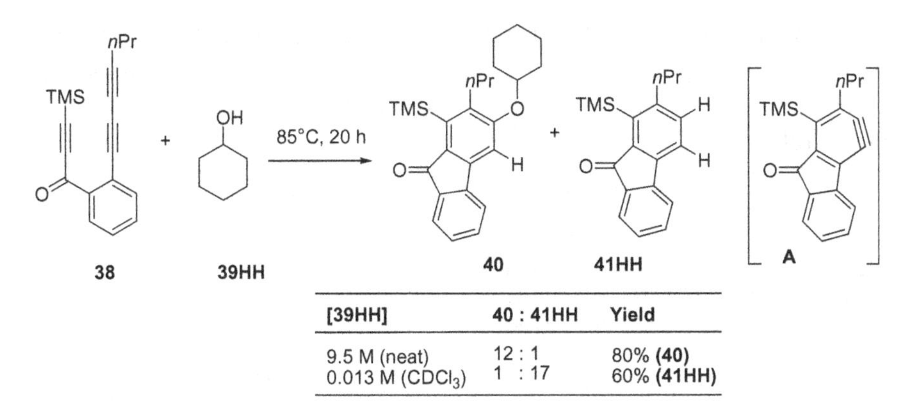

[39HH]	40 : 41HH	Yield
9.5 M (neat)	12 : 1	80% **(40)**
0.013 M ($CDCl_3$)	1 : 17	60% **(41HH)**

Scheme 4.50

Both the ether **40** and the reduced benzenoid **41HH** were formed via the benzyne intermediate **A** of distorted geometry: the ether was the alcohol addition product, and the reduced benzene was the dihydrogen addition product. Interestingly, equimolar amounts of cyclohexanone and **41HH** were formed when the reaction was conducted in $CDCl_3$.

To identify the sites in cyclohexanol from which the two hydrogen atoms were transferred, a series of deuterium-labeled cyclohexanol isotopologues were prepared (Scheme 4.51). Cyclohexanol-2,2,6,6-d_4 **39D4** was obtained by $LiAlH_4$ reduction of cyclohexanone-2,2,6,6-d_4, which was available via base-catalyzed H/D exchange reaction of cyclohexanone with heavy water.[42] H/D exchange reaction of ordinary cyclohexanol with methanol-d_4 gave cyclohexanol-OD **39-OD**, and lithium aluminum deuteride reduction of cyclohexanone afforded cyclohexanol-C1-d **39DH**.[43]

Na_2CO_3, D_2O reflux, overnight, repeat 4 times → $LiAlH_4$, Et_2O (72%) → **39D4** (98% D)

39HH → CD_3OD, repeat → **39–OD** (95% D)

$LiAlD_4$, Et_2O (77%) → **39DH**

Scheme 4.51 Deuterium-labeled substrates.

With the reagents in hand, reactions were run under standard conditions (Scheme 4.52). With 1.5 equivalents of 2,2,6,6-tetradeuteriocyclohexanol **39D4**, the fluorenone **41HH** did not have any (<5%) detectable deuterium.

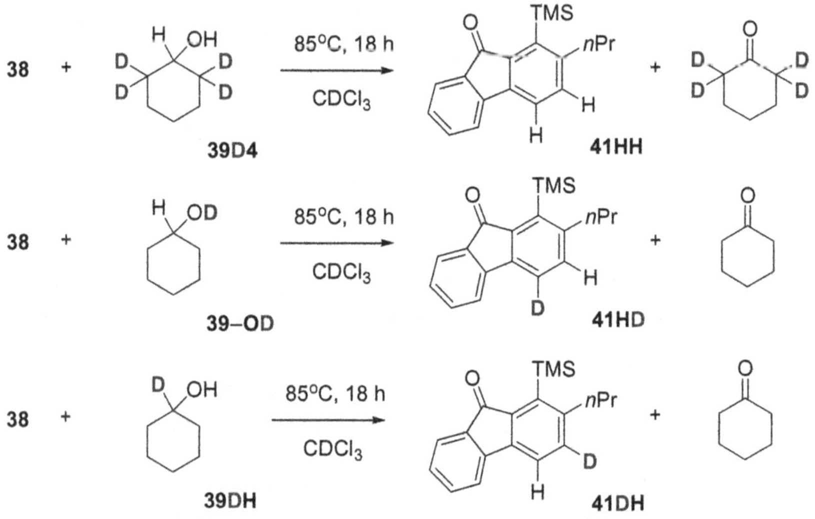

Scheme 4.52 Reactions of **38** with cyclohexanol isotopolgues.

This result suggests that the hydrogen did not come from the C-2. Next the reaction was run with cyclohexanol-OD **39-OD**: this time monodeuterated product **41HD** was formed. Curiously enough, a regioisomeric product **41DH** was obtained when the reaction was performed with cyclohexanol-1-d **39DH**. These results indicate that the hydrogens came from the alcohol OH and the carbinol methine hydrogen atom.

Finally, a KIE for the reaction of **38** with **39DH** was computed using benzyne **A** where the *n*-propyl group was replaced by a methyl group. The KIE of 1.97 for the C–H transfer process was calculated. To determine the experimental KIE, a reaction was then run using a mixture of ordinary cyclohexanol and deuterated cyclohexanol **39DH** (Scheme 4.53). The KIE was determined to be 2.0. The data is consistent with an expected early transition state for this H_2-transfer reaction.

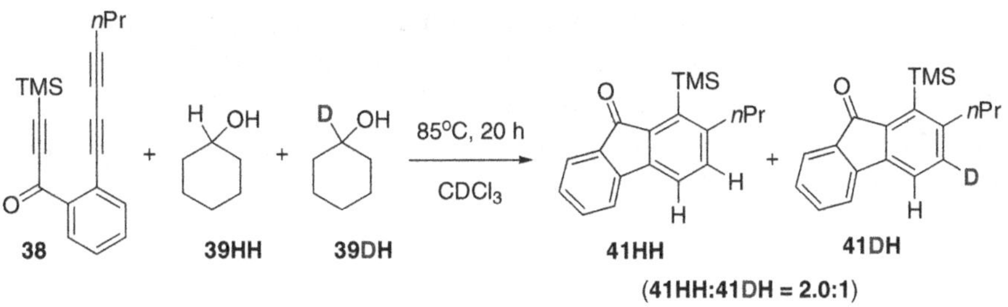

Scheme 4.53 Kinetic isotope effect.

The reaction can now be depicted to go through a transition state **TS A** (Scheme 4.54). This model involves concerted transfer of both hydrogen atoms with the deuteride adding to the more electrophilic carbon (C-3) of the benzyne intermediate. In other words, the carbinol C–H (C–D) adds to the benzyne in hydride-like and the O–H in proton-like fashion.

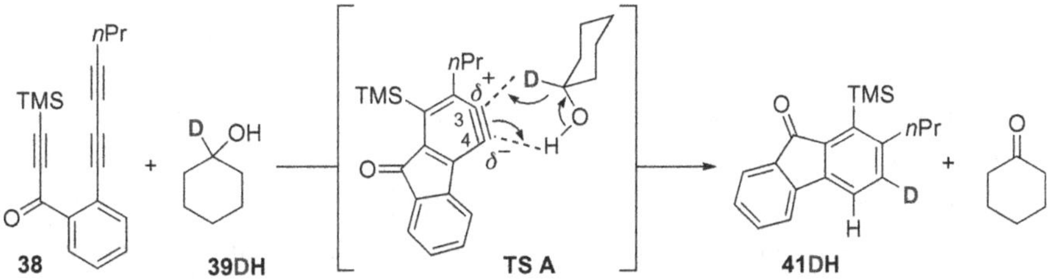

Scheme 4.54 Mechanism.

4.3.5 Reaction #15: Palladium-Catalyzed Fluorination of Aryl Triflates

The aryl triflate **42** underwent fluorination reaction when it was heated with cesium fluoride, palladium catalyst, and the biaryl phosphine ligand *t*BuBrettPhos (**A**) to provide in 70% yield as a mixture of the fluorobenzenes (**43a** and **43b**) in a 1.5:1 ratio (Scheme 4.55).[44]

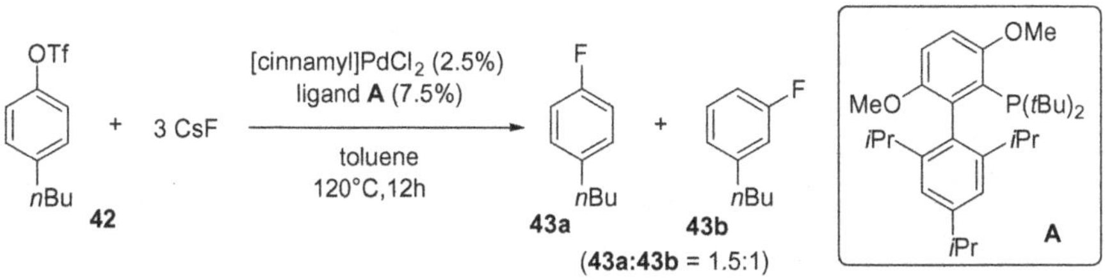

Scheme 4.55

A plausible mechanism for the formation of regioisomer **43b** is ortho-deprotonation of a catalytic intermediate **44** to generate a Pd–aryne species **45**, which would provide a mixture of regioisomers **43a/43b** via the C–F reductive elimination of the Pd complexes **46a/46b** (Scheme 4.56).

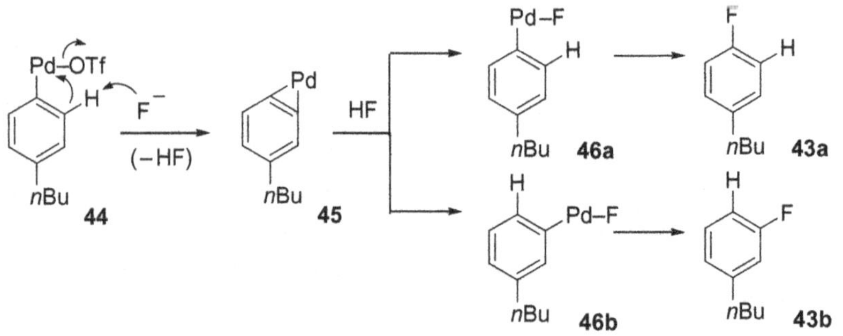

Scheme 4.56 Mechanism.

To further understand the reaction mechanism, a research group led by Professor Buchwald at Massachusetts Institute of Technology carried out a reaction using a deuterium-labeled reagent. In the process of aryne formation, HF would be formed. If a deuterium-labeled compound was added to the reaction media that would generate deuterium fluoride (DF) in situ, the products would have deuterium incorporated in the molecule. To this end, the fluorination reaction was repeated with one equivalent of *t*-butanol-OD added to the reaction mixture (Scheme 4.57).

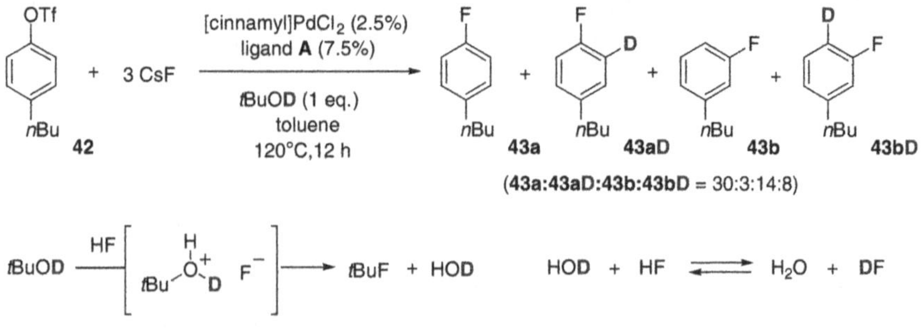

Scheme 4.57 Reactions using tBuOD.

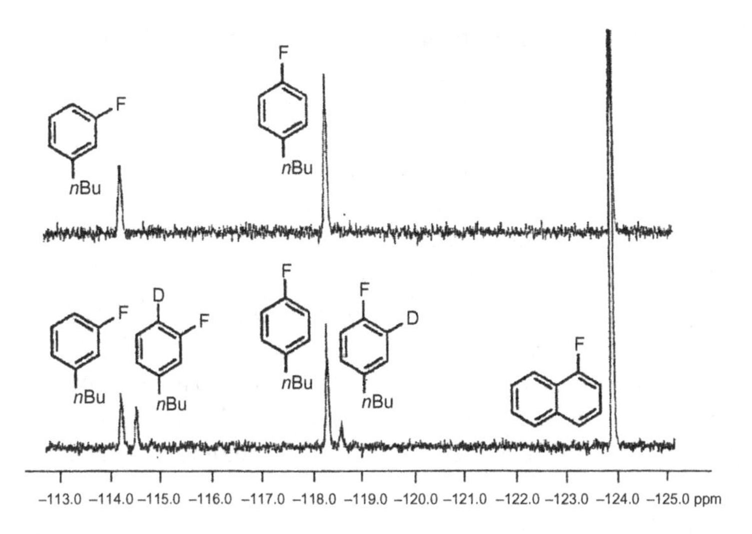

Figure 4.8 *^{19}F NMR spectra of the products: top—reaction without* t*BuOD; bottom—reaction with* t*BuOD (1-fluoronaphthalene was used as a reference).* Source: Reprinted from: Milner PJ, Kinzel T, Zhang Y, Buchwald SL. *J Am Chem Soc* 2014;**136**:15757. Copyright 2014 American Chemical Society.

DF would be formed from *t*-butanol-OD and HF, which would be generated during the formation of the palladium−aryne complex **45**. If any of the products came from the direct C−F reductive elimination pathway, there would be no deuterium incorporation in the product. It turns out that 20% of the products contained deuterium as determined by GC/MS analysis. Moreover, four aryl fluoride products (**43a/43aD/43b/43bD**) were seen in the ^{19}F NMR spectrum (Fig. 4.8). The results suggest that the Pd−aryne pathway via the formation of **45** is a competing process during the catalytic fluorination of **42**.

4.3.6 Reaction #16: Ni-Catalyzed Coupling of *N*-Tosylaziridines with Alkylzinc Bromides

A variety of aliphatic *N*-tosylaziridines (**47a**, **47b**) underwent a coupling reaction with 3-methylbutylzinc bromide to afford products (**48a**, **48b**) in the presence of organo-nickel catalyst preformed from nickel chloride and 3,4,7,8-tetramethyl-1,10-phenanthroline (Me_4phen) (Scheme 4.58).[45]

Ts N R + ZnBr (3 eq) → Me_4phen/$NiCl_2$ (1.25:1) (5 mol%), DCE/THF (2:3), 26 °C, 18 h → Ts NH R

47a, R = $Ph(CH_2)_2$
47b, R = nC_4H_9

94%, [**48a**, R = $Ph(CH_2)_2$]
96%, [**48b**, R = nC_4H_9]

Scheme 4.58

Not only did the reaction provides the products in excellent yields, but also the regioselectivity was outstanding: the C−C bond formation occurred at the less hindered side by a factor of >20:1.

To examine the stereochemical course of the reaction, Professor Jamison and coworkers at Massachusetts Institute of Technology carried out a mechanistic study using a deuterium-labeled aziridine **49D** (Scheme 4.59). The alkyne **50** was treated with Schwartz's reagent. The resulting zirconium intermediate was then quenched with heavy water to give the monodeuterated alkene, which underwent epoxidation to afford the oxirane **51D**. Nucleophilic substitution by tosylhydrazine produced the alcohol **52D**, which was subjected to the Mitsunobu conditions to give the requisite aziridine **49D** in satisfactory deuterium enrichment.

MeO
50
1. Cp_2ZrHCl
THF, 21°C;
then D_2O
2. m-CPBA
CH_2Cl_2, rt
(63%, 2 steps)
MeO
O
D
51D
$TsNH_2$ (2 eq)
$BnNEt_3Cl$ (0.1 eq)
K_2CO_3 (0.1 eq)
p-Dioxane
90°C, 5 h
MeO
OH
D
52D NHTs
DEAD (1.1 eq)
PPh_3 (1.1 eq)
THF
0°C to rt,12 h
MeO
Ts
N
D
49D (>95% D)

*Scheme 4.59 Synthesis of deuterium-labeled aziridine **49D**.*

A coupling reaction of deuterium-labeled aziridine **49D** was repeated using organozinc reagent **53** (Scheme 4.60). The expected product **54D** was formed in 75% yield with complete inversion at the deuterium-attached carbon.

MeO
Ts
N
D
49D (>95% D)
+
O
O
ZnBr
53 (3 eq)
$Me_4phen/NiCl_2$
(1.25:1) (5 mol%)
LiCl (3 eq)
DCE/DMA (1:1)
35°C, 24 h
MeO
Ts
NH
O
O
54D D

*Scheme 4.60 Reactions with **49D**.*

Based on the results, a catalytic cycle was proposed (Scheme 4.61). The active Ni(0) catalyst **A** undergoes oxidative insertion into the less hindered carbon of the aziridine via an S_N2-type mechanism to form the nickel-containing complex **B**. Subsequent ring-opening reaction with the organozinc reagent gives intermediate **C**, which releases the product and the catalyst **A** via reductive elimination.

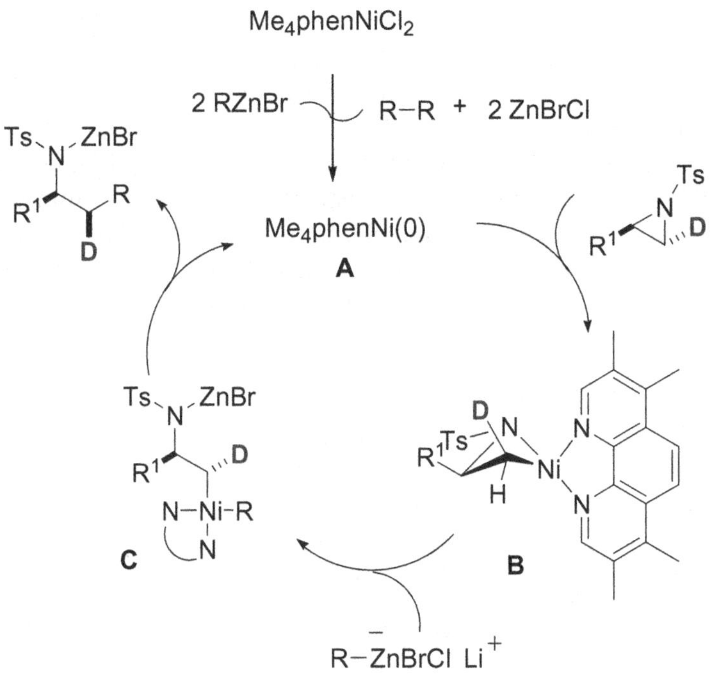

Scheme 4.61 Mechanism. Source: Reprinted from: Jensen KL, Standley EA, Jamison TF. *J Am Chem Soc* 2014;**136**:11145. Copyright 2014 American Chemical Society.

4.3.7 Reaction #17: Reaction of Arylboronic Acids with Alkenyl Alcohols

In 2013, Professor Sigman and coworkers at the University of Utah reported an enantioselective Heck-type reaction of acyclic alkenol substrates with aryl boronic acids. When three equivalents of arylboronic acid **55** were stirred with the homoallylic alcohol **56** and molecular sieves in the presence of chiral ligand **A**, palladium and copper catalysts under an oxygen atmosphere, the γ-addition product **57a** was isolated in 82% yield with 98% ee. The β-addition product **57b** was formed as a minor product (Scheme 4.62).[46]

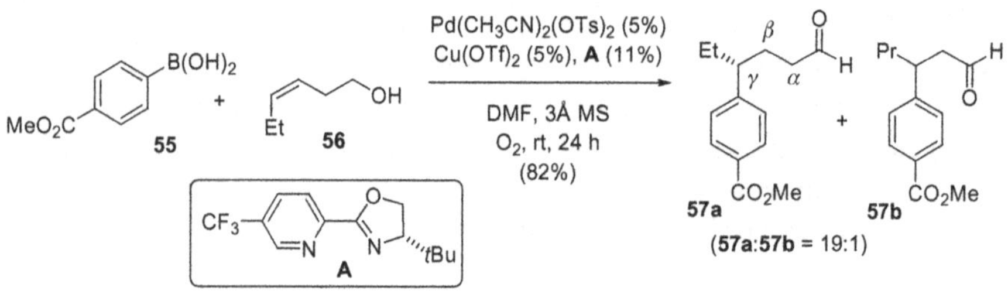

Scheme 4.62

The authors proposed the following mechanism (Scheme 4.63). A migratory insertion of the palladium–aryl species to the alkenol **56** would give intermediate **P3**, which upon β-hydride elimination produces **P3a**. Next, the Pd-bound alkenol **P3a** undergoes another round of migratory insertion/β-hydride elimination to give **P2b** via the intermediate **P2**. A migratory insertion of the palladium–hydride species in **P2b** would yield intermediate **P1**, which then releases the γ-aryl substituted aldehyde **57a**.

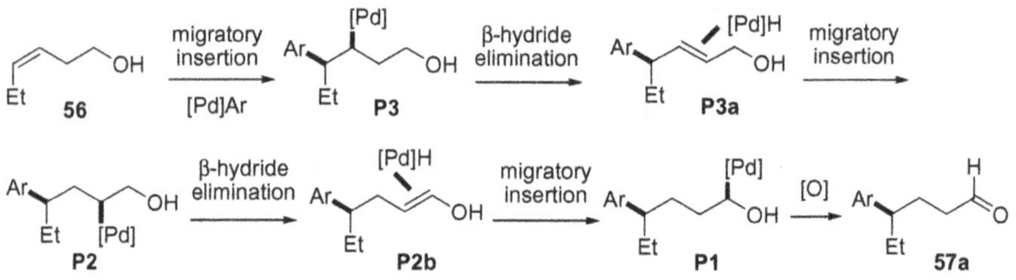

Scheme 4.63 Mechanism.

During the process, the palladium effectively walks along the carbon chain toward the hydroxy group: C3→C2→C1 (from C3 in **P3** to C2 in **P2**, then finally to C1 in **P1**). One of the questions the researchers asked was whether or not the palladium walked in only one direction towards the alcohol. In other words, was the walking back and forth random?

To investigate the reversibility of the relay process, computational studies were performed: the barriers between the β-hydride elimination and migratory insertions are relatively low, suggesting that the palladium catalyst could migrate both away from and toward the alcohol before releasing the product.

To experimentally determine the reversibility of the relay process, the alkenol-d2 substrate **56D** was synthesized and then subjected to the oxidative redox-relay Heck reaction conditions (Scheme 4.64).[47] Thus stereoselective reduction of 3-hexyn-1-ol with sodium borodeuteride catalyzed by nickel acetate under an atmosphere of deuterium gas afforded the alkenol **56D** in 48% yield and 89% isotopic purity.[48] When the reaction was carried out under usual conditions, expected product **57aD** was formed in 68% yield. Interestingly both deuterium were found at the β- and γ-carbon without any loss.

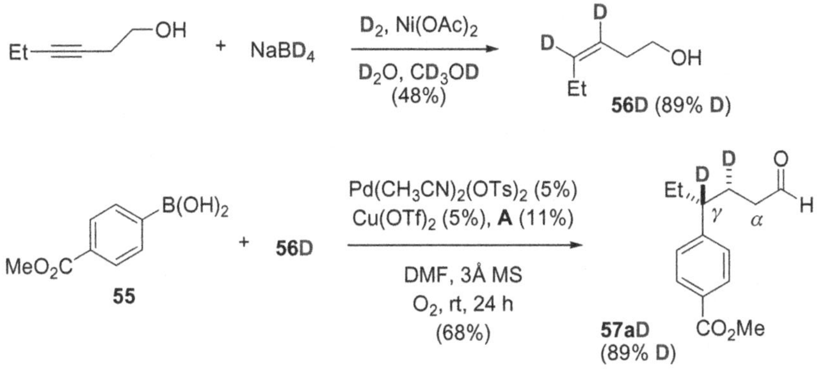

*Scheme 4.64 Deuterium-labeled substrate **56D**.*

If there were any reversible walk during the process, one would expect to isolate the isotopomer **57aD'** that contains deuterium at the α-position. The absence of **57aD'**, confirmed by experiment, suggests that the palladium walk is one direction only and towards the alcohol (Scheme 4.65).

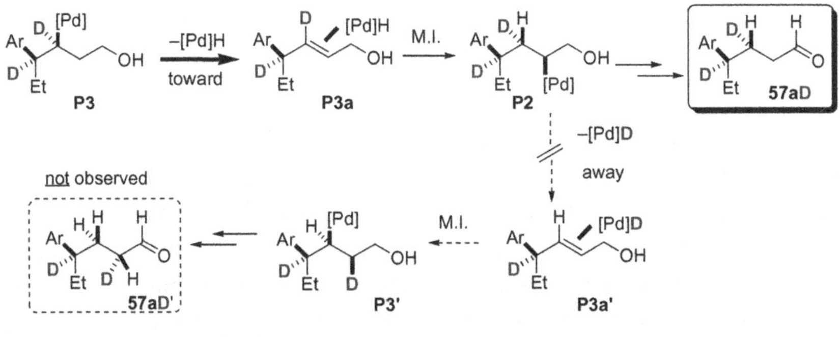

Scheme 4.65 Mechanism—It is towards, not away.

Another deuterium-labeled substrate **58D** was synthesized to determine the reaction pathway for the last step of the Heck reaction, the product-forming step (Scheme 4.66). Reduction of methyl (*Z*)-hex-3-enoate with lithium aluminum deuteride gave the homoallylic alcohol **58D** of greater than 95% isotopic enrichment in 39% yield. The reaction of boronic acid **55** with the alcohol **58D** under usual conditions afforded the product **59aD**, which has deuterium attached both at C-1 and C-2 positions.

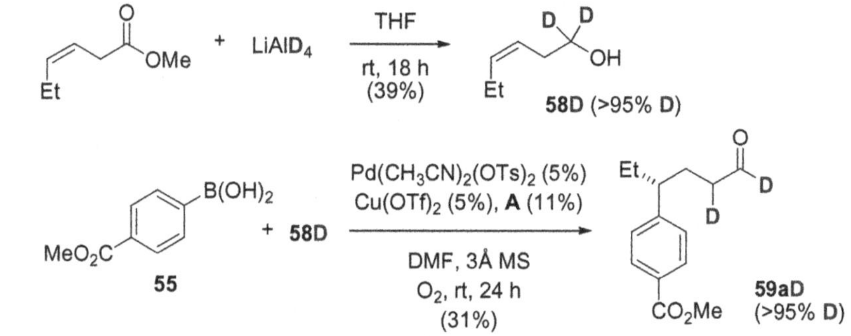

*Scheme 4.66 Deuterium-labeled substrate **58D**.*

This result suggests that the C-2 palladium alkyl intermediate **P2D** yields the palladium complex **E**, which upon migratory insertion produces the C-1 palladium alkyl intermediate **P1D**. Oxidative deprotonation would then finally deliver the aldehyde **59aD** (Scheme 4.67).

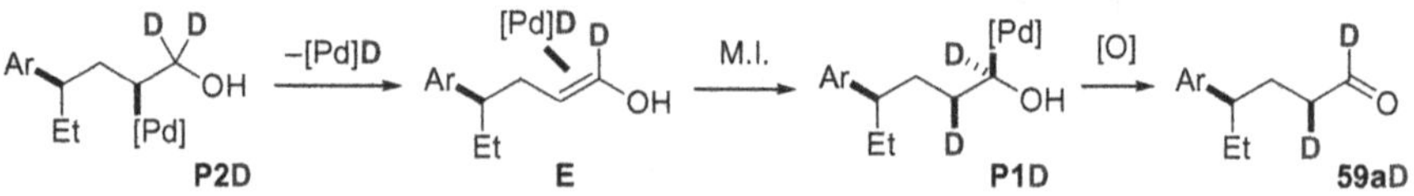

Scheme 4.67 Mechanism.

4.3.8 Reaction #18: Iodination of Resorcinol

Although bromination of benzene and benzene-d_6 proceeds at identical rates, a substantial KIE for the iodination of phenol with molecular iodine was reported with an observed k_H/k_D up to 6.3.[49] Professor Jung and coworkers at the University of Southern California developed an efficient laboratory experiment to demonstrate the kinetic difference between ordinary resorcinol and deuterated resorcinol in the iodination. First, H/D exchange of resorcinol was carried out in acidic D_2O solution (Scheme 4.68).[50]

HO OH dilute H_2SO_4, D_2O, reflux, 30 min → DO OD D D D

Scheme 4.68 H/D exchange reaction.

The extent of deuteration was conveniently followed by ^1H NMR by integrating the proton at the C-5 position and comparing it to any remaining proton signals (Fig. 4.9).

With deuterium-labeled resorcinol in hand, iodination experiment was conducted to determine the KIE (Scheme 4.69). Before addition of iodine, both reaction mixtures containing ordinary resorcinol and deuterated resorcinol were colorless. After addition, both turned dark orange red. The iodine color disappeared within 5 minutes for the ordinary resorcinol, whereas it took nearly 30 minutes for the deuterated resorcinol. To determine this rate difference quantitatively, the reaction was monitored by UV–VIS spectrometer: a value of $k_H/k_D = 3.4$ was obtained.

HO OH I_2, NaI, H_2O, EtOH, k_H → HO OH I

DO OD D D D I_2, NaI, D_2O, EtOH, k_D → DO OD D D I

$\frac{k_H}{k_D} = 3.4$

Scheme 4.69 Kinetic isotope effect.

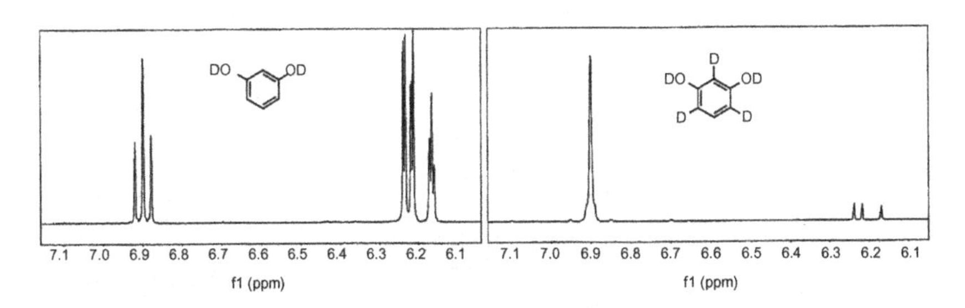

Figure 4.9 ^1H NMR spectra of ordinary and deuterium-labeled resorcinols. Source: Reprinted from: Giles R, Kim I, Chao WE, Moore J, Jung KW. *J Chem Ed* 2014;**91**:1220. Copyright 2014 American Chemical Society.

Based on the results, a simplified mechanism of iodination was suggested (Scheme 4.70). The iodine addition to the resorcinol occurs fast and in a reversible manner to give intermediate **A**. The second step of the reaction is deprotonation by water. The KIE indicates that the C–H bond breaking of the intermediate **A** is part of the rate-determining step.

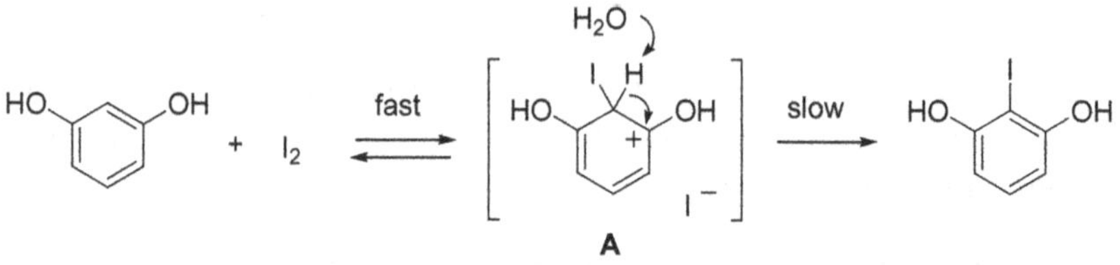

Scheme 4.70 Mechanism.

4.3.9 Reaction #19: Iridium-Catalyzed Asymmetric Hydroheteroarylation of Norbornene

Christo Sevov and Professor Hartwig at the University of California, Berkeley, reported that the C–H bonds of a variety of heterocycles underwent a highly enantioselective addition reaction to norbornene.[51] For example, the C-2 alkylated product **60a** was obtained with indole in 93% yield and 96% ee, and with pyrrole, the corresponding product **60b** was obtained in 61% yield and 93% ee (Scheme 4.71). The reaction was mediated by the catalyst formed by iridium metal complex, $[Ir(coe)_2Cl]_2$ (coe = cyclooctene) and a chiral phosphine ligand, (*S*)-DTBM-Segphos (Ar = 3,5-di-*tert*-butyl-4-methoxyphenyl).

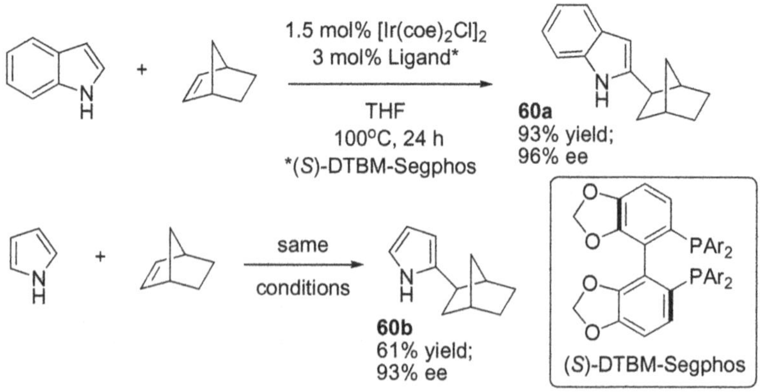

Scheme 4.71

To understand the mechanism of the reaction, a deuterium-labeled indole compound **61D** was prepared (Scheme 4.72). Deprotonation of the sulfonamide **62** with *n*-butyllithium followed by quenching with methanol-OD provided the sulfonamide-d_2 **62D2**, which upon saponification afforded the monodeuterated indole **61D** of high deuterium content.

n-BuLi, THF
then CH_3OD
(95%)
2 M NaOH
CH_3OH
90°C, 21 h
(68%)
62
62D2
61D (98% D)

Scheme 4.72 Preparation of deuterium-labeled substrate.

With 2-deuterio-indole **61D** in hand, the asymmetric reaction was repeated under standard conditions (Scheme 4.73). It was determined that the indole moiety and deuterium in **60aD** are *syn* to each other. This *syn*-stereochemistry implies that the hydroheteroarylation reaction occurs by a mechanism involving olefin insertion into an Ir–C or Ir–H bond.

1 mol% $[Ir(coe)_2Cl]_2$
2 mol% Ligand*
80°C, THF
*(*S*)-DTBM-Segphos
61D
60aD
(85% yield; >95% D)

*Scheme 4.73 Reactions with **61D**.*

Another key piece of information was obtained when the reactions were run with ordinary indole and deuterated indole **61D**: a KIE of 1.4 was measured (Scheme 4.74). The relatively small number indicates that the C–H (C–D) bond breaking is not the rate-determining step of the catalytic cycle.

Ir. catalyst
80°C, THF
k_H
60a
Ir. catalyst
80°C, THF
k_D
61D
60aD
$\frac{k_H}{k_D} = 1.4$

Scheme 4.74 Kinetic isotope effect.

Based on NMR studies and deuterium-labeling experiments, a mechanism of the reaction was proposed (Scheme 4.75). The C–D bond of the heterocycle adds to the active catalyst **A**. This step is rapid and possibly reversible. Addition of norbornene to **B** gives intermediate **C**, from which aryl group migrates to norbornene. Reductive elimination delivers the product and regenerates the catalyst **A**.

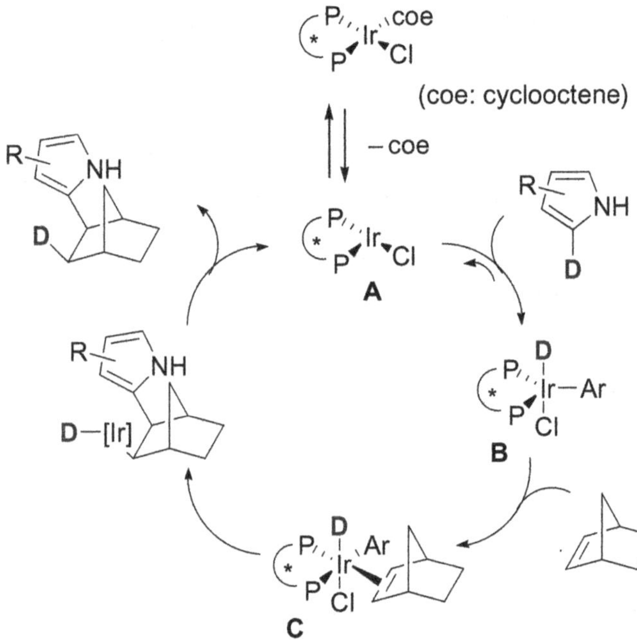

Scheme 4.75 Mechanism. Source: Reprinted from: Sevov CS, Hartwig JF. *J Am Chem Soc* 2013;**135**:2116. Copyright 2013 American Chemical Society.

4.3.10 Reaction #20: Copper(I)/TEMPO-Catalyzed Aerobic Oxidation of Alcohols

Professor Stahl and her research team of the University of Wisconsin at Madison developed a very mild reaction protocol, by which an array of primary alcohols was obtained in high yields using oxygen and a catalytic amount of Cu(I), 2,2,6,6-tetramethyl-1-piperidinyloxyl (TEMPO), 2,2'-bipyridyl (bpy), and *N*-methylimidazole (NMI) (Scheme 4.76).[52]

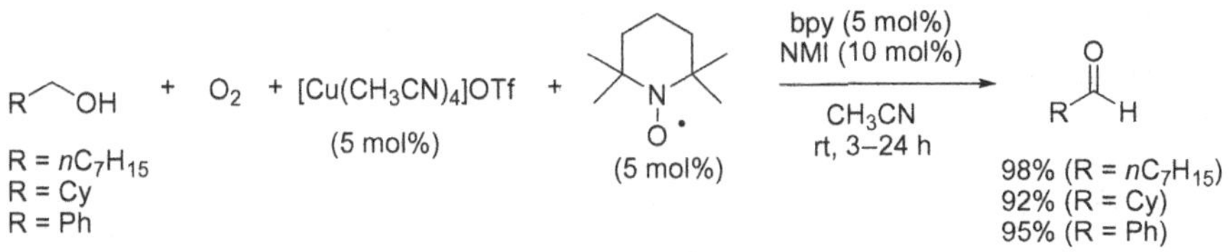

Scheme 4.76

To understand the reaction mechanism, extensive studies were carried out that included deuterium KIE. Two substrates of high deuterium enrichment were prepared by $LiAlD_4$ reduction for the study (Scheme 4.77).

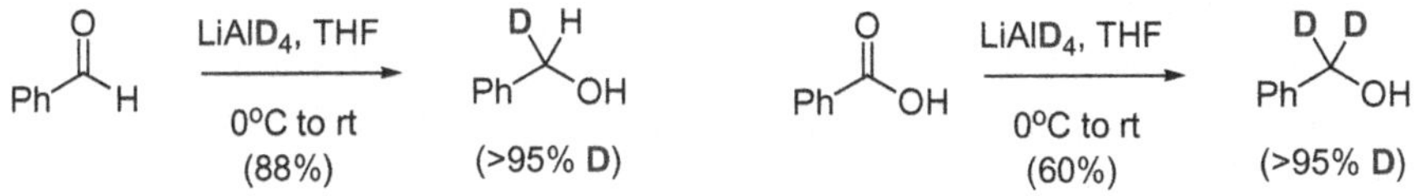

Scheme 4.77

When oxidation was run on benzylalcohol-d_1, a substantial KIE was observed (Scheme 4.78, A). Another experiment was repeated using ordinary benzylacohol and benzylalcohol-d_2 in a separate reaction vessel (Scheme 4.78, B). Surprisingly, no KIE was observed leading to a conclusion that C–H cleavage is not the rate-determining step of the reaction.[53]

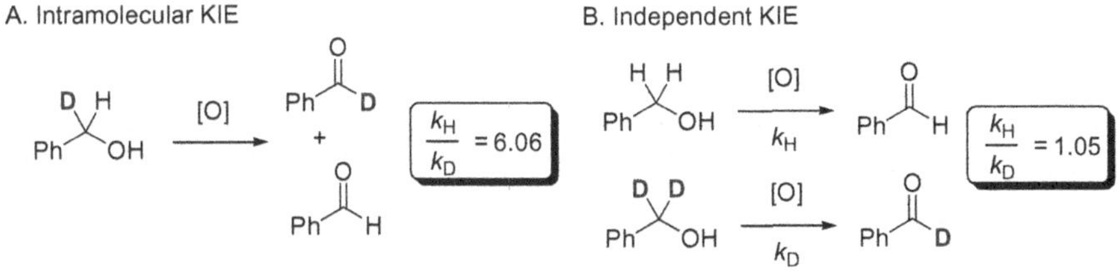

Scheme 4.78 Kinetic isotope effect.

4.3.11 Reaction #21: Total Synthesis of *N*-Methylwelwitindoline C Isonitrile

In the total synthesis of *N*-methylwelwitindoline C isonitrile, an intramolecular nitrene insertion reaction of the carbamate **63** was employed to construct the fused oxazolidinone (Scheme 4.79).[54] Unfortunately, the desired product **64** was obtained in only 33% yield. The major byproduct was the ketone **65**.

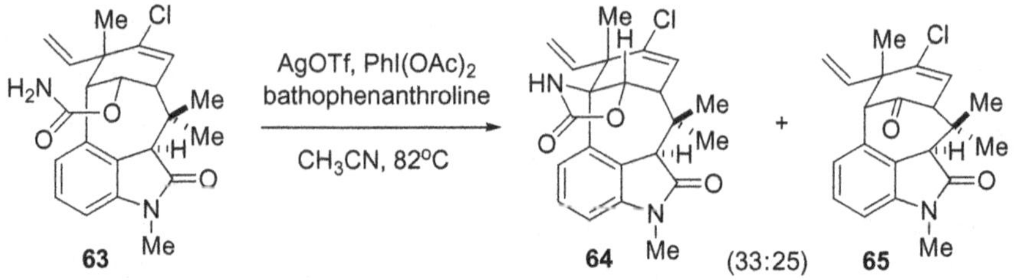

Scheme 4.79

In an attempt to minimize the formation of the ketone byproduct, Professor Garg and his team at the University of California, Los Angeles, employed a rather unusual tactic: the deuterium KIE. For this, a deuterium-labeled substrate **63D** was prepared (Scheme 4.80). A stereoselective reduction of ketone **65** with super-deuteride ($LiEt_3BD$) in THF gave alcohol **66D**, which underwent carbamoylation to afford the carbamate **63D** in quantitative yield.

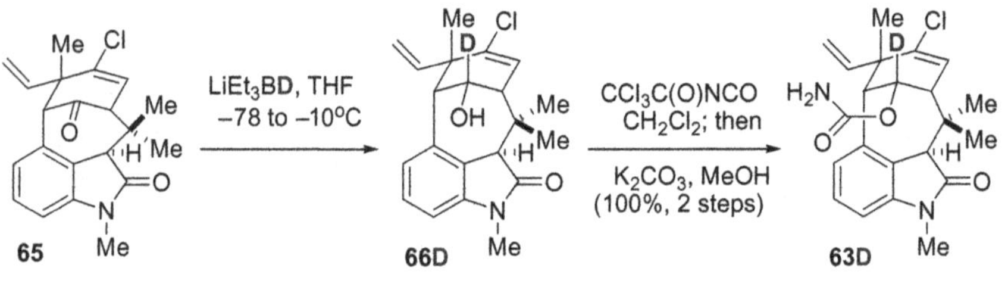

Scheme 4.80 Deuterium-labeled carbamate.

When the deuterium-labeled carbamate **63D** was subjected to the usual conditions, the desired product **64D** was formed in good yield, thanks in large part to the deuterium (Scheme 4.81).

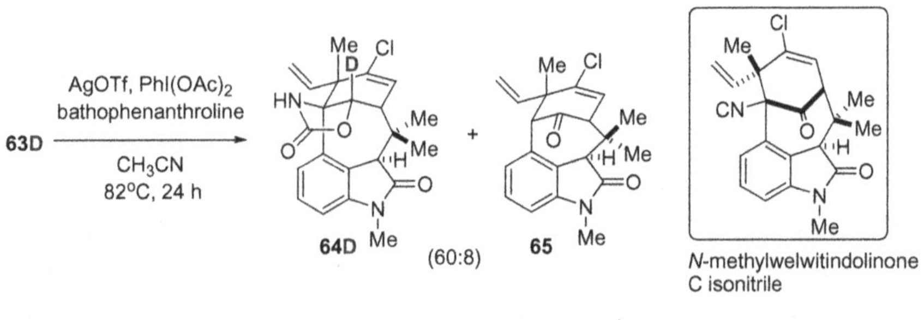

Scheme 4.81 Deuterium increases yield.

Professor Amos B. Smith III of the University of Pennsylvania commented on the work: "Garg's synthesis is an extraordinary innovation and one of the most elegant examples of C–H functionalization to date in total synthesis."[55]

4.3.12 Reaction #22: Total Synthesis of Norzoanthamine

Another fine example of deuterium applications in synthetic organic chemistry was showcased in the total synthesis of an alkaloid natural product norzoanthamine. Miyashita and coworkers encountered a problem converting the methyl ketone **67** to the alkyne **68** (Scheme 4.82).[56] A significant amount of the dihydropyran byproduct **69** was formed as a result of the [1,5]-hydride shift.

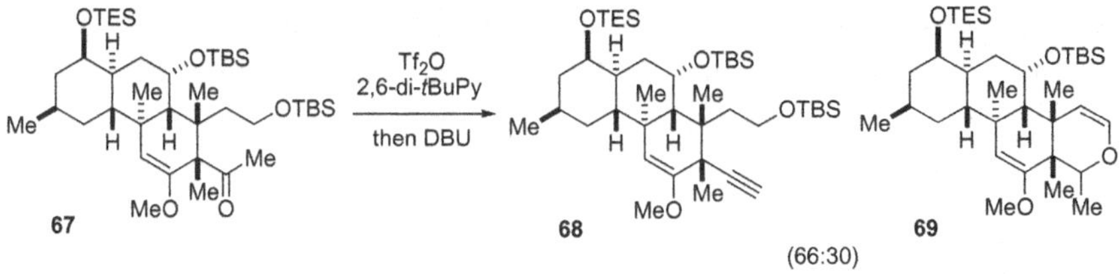

Scheme 4.82

In an effort to enhance the selectivity, a deuterium substrate **67D** was prepared (Scheme 4.83). Wittig reaction of lactol **70** with triphenylphosphonium bromide-methyl-d_3 in the presence of KHMDS gave the alcohol **71** after hydroboration of the alkene. After five steps, the deuterium-labeled substrate **67D** was ready.

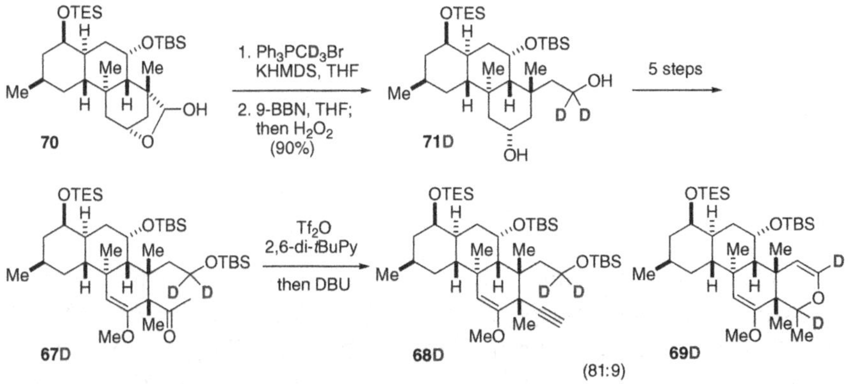

Scheme 4.83 Deuterium improves selectivity.

The methyl ketone **67D** was treated with triflic anhydride in the presence of 2,6-di-*t*-butylpyridine in 1,2-dichloroethane at rt for 3 hours. After that, the reaction mixture was heated with DBU base at 80°C for 5 hours. Surprisingly, the desired alkyne **68D** was formed in 81% yield. The selectivity of alkyne formation over the cyclization event more than quadrupled to 9:1 from mere 2:1 by simply utilizing deuterium.

4.3.13 Reaction #23: Scandium-Catalyzed C–H Functionalization

When the benzylidene barbituric acids (**72a**, **72b**) were heated with a catalytic amount of scandium triflate in 1,2-dichloroethane, tetraline derivatives (**73a**, **73b**) were obtained in excellent yields (Scheme 4.84).[57] The reaction is presumed to go through a carbocation intermediate **A**, which is formed via an intramolecular [1,5]-hydride shift.

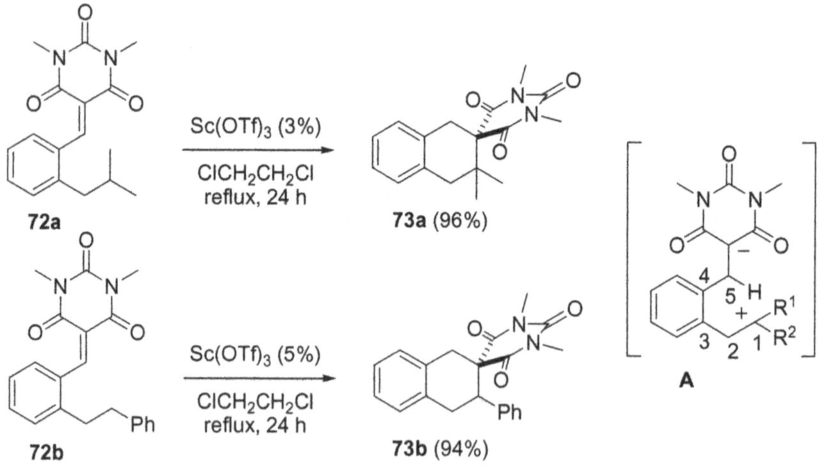

72a — $Sc(OTf)_3$ (3%), $ClCH_2CH_2Cl$, reflux, 24 h → 73a (96%)

72b — $Sc(OTf)_3$ (5%), $ClCH_2CH_2Cl$, reflux, 24 h → 73b (94%)

A

Scheme 4.84

To check the reversibility of the so-called internal redox process, Professor Akiyama, a recipient of the 2016 Cope Scholar Award, and coworkers prepared a deuterium-labeled substrate **72bD** (Scheme 4.85). Reduction of 2-phenylethylbenzoic acid with $LiAlD_4$ followed by MnO_2 oxidation gave 2-phenylethylbenzaldehyde-d_1 **74D** in quantitative yield. Reaction of the benzaldehyde derivative **74D** with 1,3-dimethylbarbituric acid (1,3-DMBA) then afforded the benzylidene **72bD**. When the bezylidene **72bD** was treated with scandium triflate, the only product formed was tetraline **73bDH**, in which deuterium was located at the benzylic position. This confirms that the [1,5]-hydride shift is irreversible.

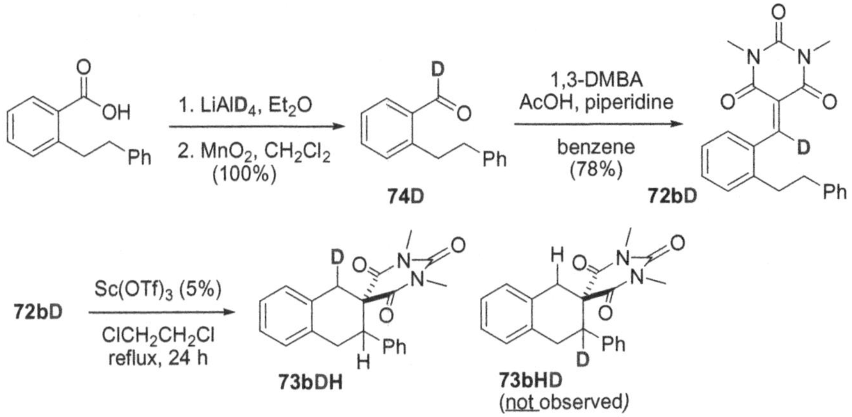

Scheme 4.85 Deuterium-labeling experiments.

Another interesting observation was made when the compound **75** was treated with scandium triflate (Scheme 4.86). Surprisingly, none of the expected seven-membered ring adduct was observed. Instead, an indane derivative **76** was formed.

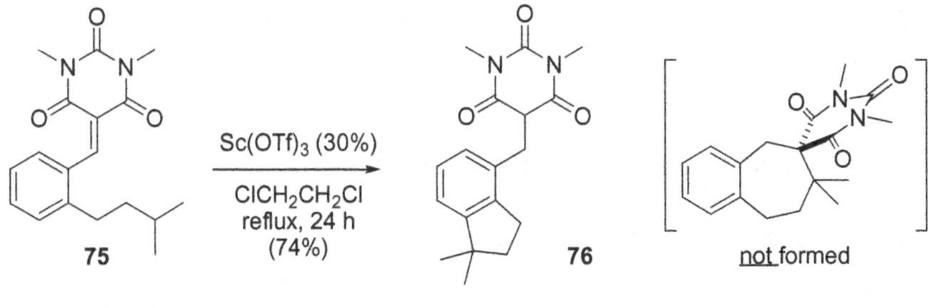

Scheme 4.86

To prove that the reaction occurs through a [1,6]-hydride shift, a deuterium-labeled substrate **75D** was prepared (Scheme 4.87). Quenching of the lithium enolate of ethyl isobutyrate with D_2O gave α-deuterated ester. Subsequent treatment with $LiAlH_4$, O-tosylation and iodide displacement gave the isobutyl iodide-d. Next C-alkylation of 2-methylbenzylalcohol with the iodide afforded the corresponding benzyl alcohol **77D**. Oxidation gave aldehyde **78D**, which underwent condensation reaction to afford the required substrate **75D**.

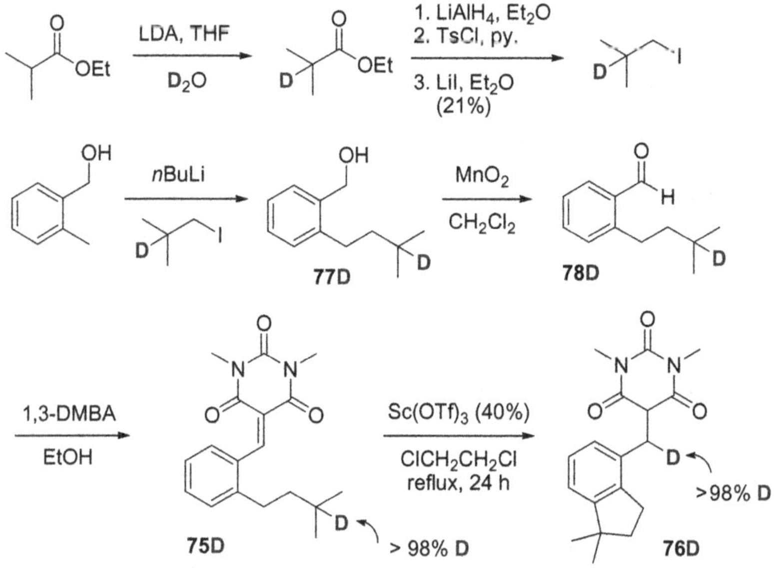

Scheme 4.87 Deuterium-labeling experiments.

When the reaction was repeated with **75D**, the product **76D** contains deuterium only at the benzylic position (Fig. 4.10). This result confirms that [1,6]-hydride (deuteride) shift has occurred during the formation of indane.

4.3.14 Reaction #24: Meinwald Rearrangement Involving 1,2-Boryl Migration

In 2011, Professor Yudin and coworker at the University of Toronto reported that a series of α-boryl aldehydes (**80a**, **80b**) could be prepared when oxiranyl *N*-methyliminodiacetyl boronates (**79a**, **79b**) are treated with BF_3 etherate (Scheme 4.88, Eq. 1).[58] In a back-to-back publication, Professor Burke and coworker at the University of Illinois at Urbana-Champaign communicated their finding that the chiral epoxide (**79c**) underwent the same rearrangement reaction in the

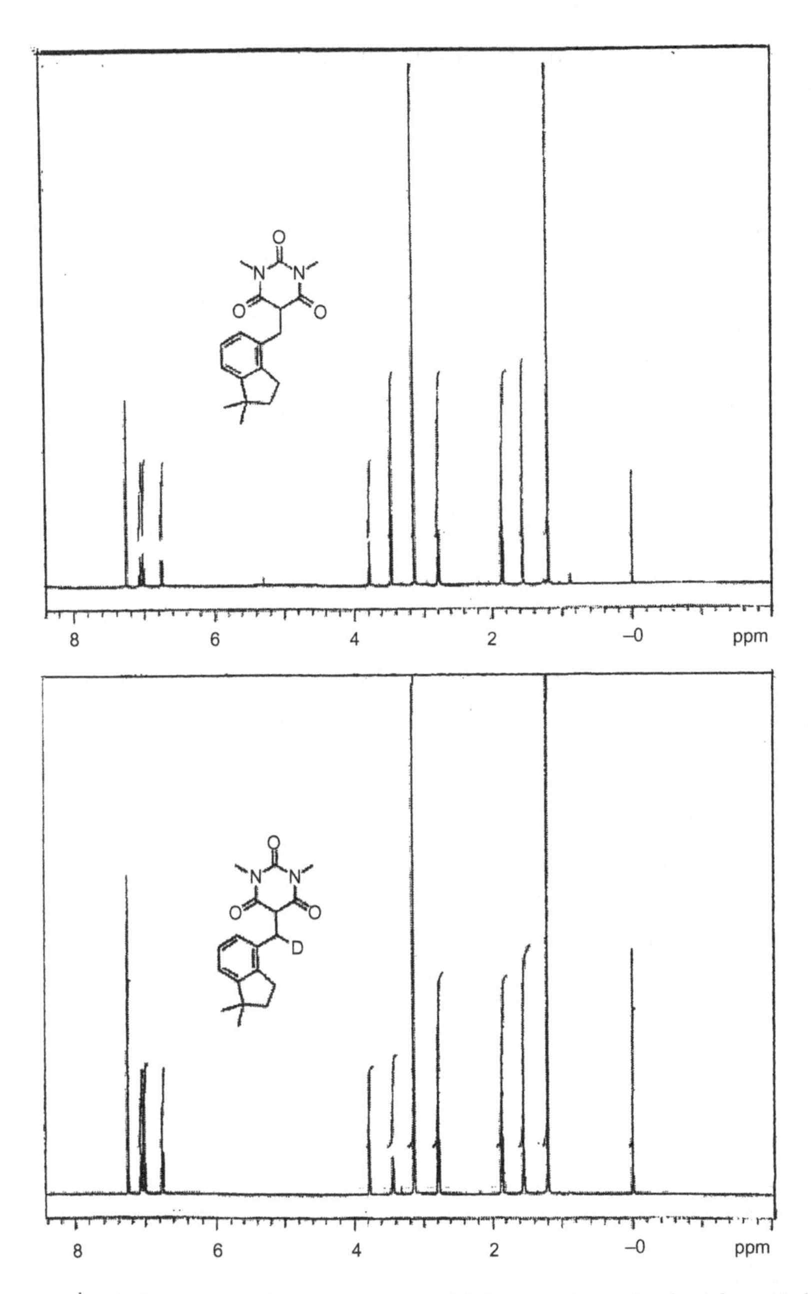

Figure 4.10 ^{1}*H NMR spectra of ordinary and deuterium-labeled indanes.* Source: Reprinted from: Mori K, Sueoka S, Akiyama T. *J Am Chem Soc* 2011;**133**:2424. Copyright 2011 American Chemical Society.

presence of magnesium perchlorate to afford the chiral aldehyde **80c** with perfect retention of chirality (Scheme 4.88, Eq. 2).[59]

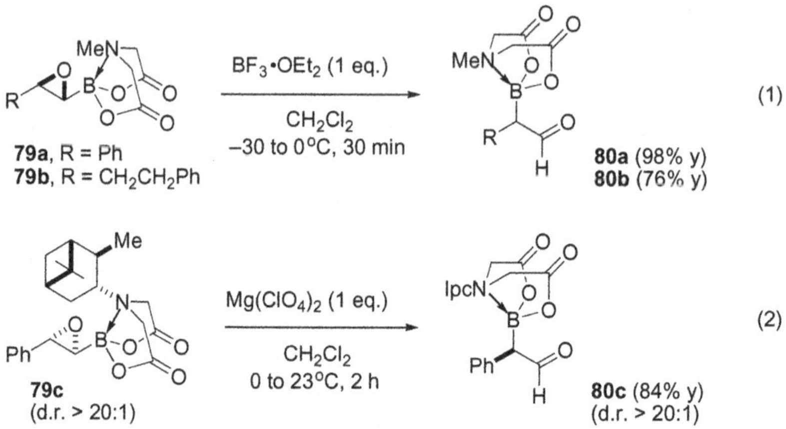

Scheme 4.88

Only the aldehyde **80a** was formed with the ketone being absent. To further investigate the regioselectivity of the reaction, Yudin's lab conducted a mechanistic study using a deuterium-labeled substrate **79aD** (Scheme 4.89). Thus treatment of phenyl acetylene with *n*-butyl lithium followed by quenching with heavy water gave phenyl acetylene-d. Hydroboration followed by hydrolysis provided deuterated boronic acid **81D**. Treatment of the boronic acid with *N*-methyliminodiacetic acid gave the corresponding boronate **82D**, epoxidation of which afforded a deuterium-labeled substrate **79aD**.

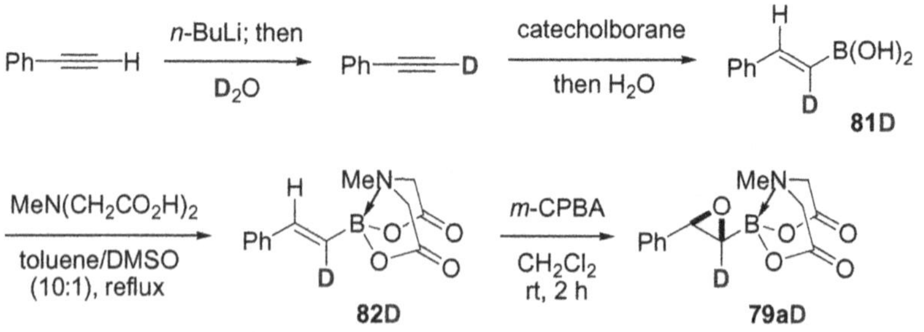

*Scheme 4.89 Synthesis of deuterium-labeled substrate **79aD**.*

When the epoxide **79aD** was subjected to the rearrangement conditions, the aldehyde **80aD** was formed that had deuterium bound to its aldehyde carbon (Scheme 4.90). The labeling experiment established that a 1,2-boryl migration occurred from the intermediate **A** with concomitant opening of the epoxide.

$BF_3 \cdot OEt_2$ (1 eq.), CH_2Cl_2, −30 to 0°C, 30 min (98%); 79aD → 80aD; A

Scheme 4.90 Deuterium-labeling experiments.

4.3.15 Reaction #25: Ring-Closing Metathesis—the Case of Nonproductive Events

Ring-closing metathesis of dienes is an efficient way to access carbocycles and heterocycles of different sizes.[60] For example, reaction of diallyl ether **83a** with the ruthenium catalyst **A** afforded the metathesis product, 2,5-dihydrofuran **84a** as the major product. The same catalyst worked very well on diethyl diallylmalonate **83b** to produce the cyclopentene **84b** in excellent yield (Scheme 4.91).

(1) 83a → 84a (>95% conv.); catalyst A (5 mol%), benzoquinone (0.1 eq), CD_2Cl_2, 40°C, 1 h

(2) 83b (EtO_2C, CO_2Et) → 84b (>95% conv.); catalyst A (1 mol%), CD_2Cl_2, 30°C, 0.5 h

A: Cl_2Ru=CHPh, PCy_3

Scheme 4.91 Ring-closing metathesis.

In an effort to increase the efficiency of the ring-closing metathesis reaction, Professor Grubbs and coworkers at California Institute of Technology needed to find out how much nonproductive events occurred during the reaction. For this purpose, deuterium-labeled substrate **83bD2** was prepared (Scheme 4.92).[61]

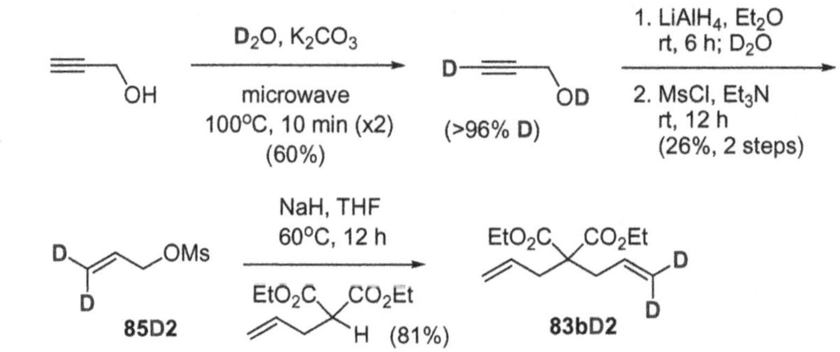

Scheme 4.92 *Synthesis of deuterium-labeled substrate.*

Propargyl alcohol was mixed with heavy water and heated in a microwave in the presence of potassium carbonate to give, after another H/D exchange reaction, propargyl alcohol-d_2 of satisfactory deuterium enrichment. LAH reduction followed by mesylation afforded the mesylate-d_2 **85D2**, C-alkylation of which then produced the desired substrate **83bD2** in 81% yield.

Ring-closing metathesis reaction was run on deuterium-labeled substrate **83bD2** using a number of ruthenium catalysts (Scheme 4.93). The nonproductive event was followed by observing the appearance of **83b** and **83bD4** with time-of-flight mass spectrometry (TOF-MS).

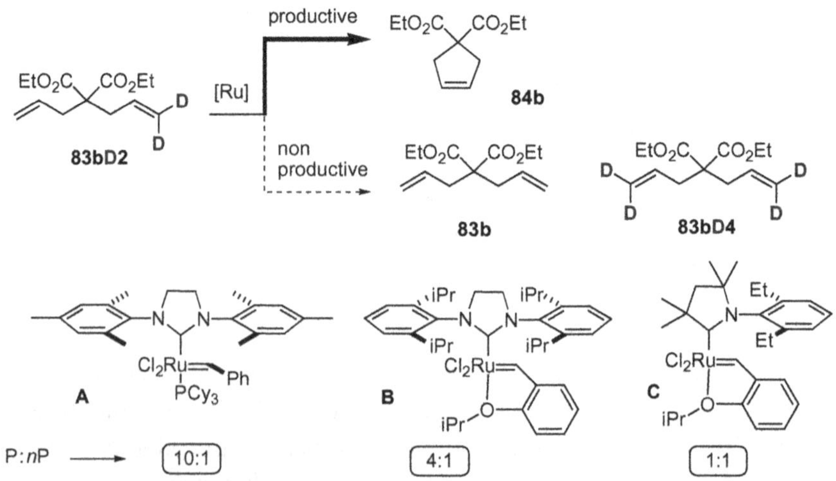

Scheme 4.93 *Productive versus nonproductive events.*

Although a relatively small amount (~10%) of nonproductive event occurred with the catalyst **A**, a nearly equal number of productive and nonproductive metathesis events happened with the catalyst **C**. The data hints that steric effects are the overriding determinants in the amount of nonproductive events.

The use of deuterium-labeled substrate **83bD2** was crucial in the study. Without deuterium, the nonproductive events could not have been detected[62]:

Robert H. Grubbs was born on February 27, 1942, in Possun Trot, KY. He received his BA and MSc degrees from the University of Florida working with Professor Merle Battiste and his PhD from Columbia University under Professor Ronald Breslow. After a postdoctoral year with Professor Jim Collman at Stanford University, he joined the faculty at Michigan State University where he started his work on olefin metathesis. In 1978, he moved to California Institute of Technology, where he is now the Victor and Elizabeth Atkins Professor of Chemistry. He was awarded a 2005 Nobel Prize in chemistry for the development of the metathesis method in organic synthesis. He shared the Prize with Yves Chauvin and Richard R. Schrock.

4.3.16 Reaction #26: Rearrangement of (*Z*)-Allylic Trichloroacetimidates to Allylic Esters

The (*Z*)-allylic trichloroacetimidates **86** undergo S_N2' displacement reaction to form allylic esters **87** in the presence of a chiral palladium complex **A** in high to excellent enantioselectivity (Scheme 4.94).[63]

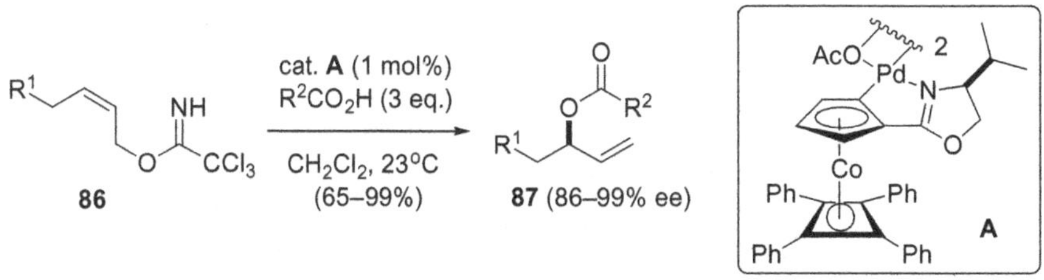

Scheme 4.94

In order to determine the geometry of the reaction, Professor Overman and coworkers at the University of California, Irvine, carried out a mechanistic study using a deuterium-labeled substrate **86D** (Scheme 4.95). Lithium aluminum deuteride reduction of ynoate **88** went smoothly to give the propargylic alcohol **89D** in high yield. Partial hydrogenation employing Lindlar's catalyst, followed by Ley oxidation afforded the aldehyde **90D**. Enantioselective Keck reduction provided the allylic alcohol **91D**, which upon treatment with trichloroacetonitrile, gave the requisite acetimidate **86D**.

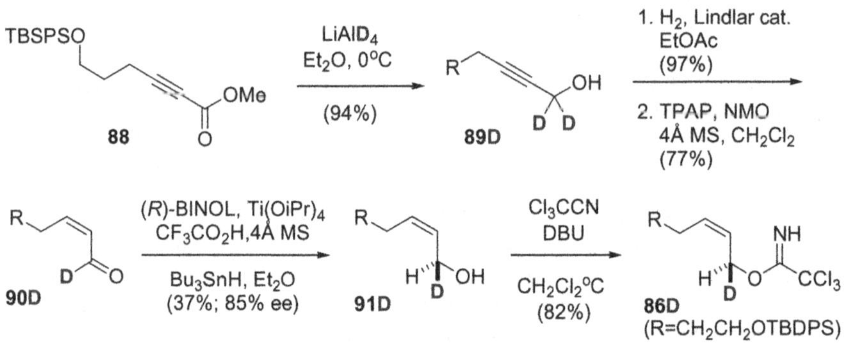

Scheme 4.95 Deuterium-labeled substrate ***86D****.*

With substrate in hand, the S_N2' reaction was performed under standard conditions (Scheme 4.96). Within experimental uncertainty, the chirality transfer was perfect: the (*S*)-imidate transformed to an (*E*)-alkene and the (*R*)-imidate became a (*Z*)-alkene. The results establish that the S_N2' reaction of the (*Z*)-allylic imidates with carboxylic acids proceeds with antarafacial stereospecificity.

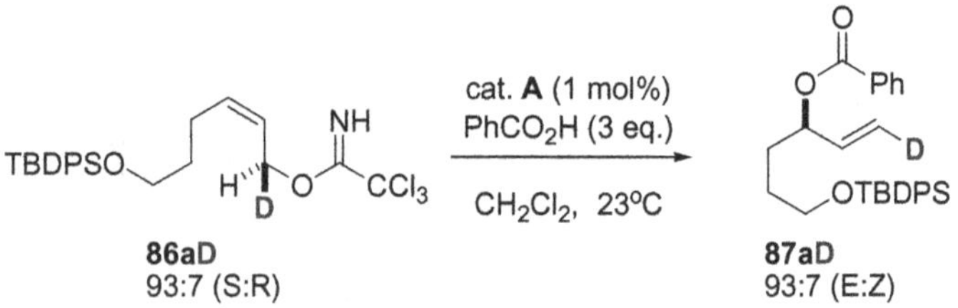

Scheme 4.96 Reaction of deuterium-labeled substrate ***86aD****.*

Based on the above results and computational studies, the following reaction mechanism was proposed (Scheme 4.97). In this mechanism, *anti*-acyloxypalladation by external nucleophilic attack of benzoic acid at the C-3 of the palladium–imidate complex generates the intermediate, which upon *syn*-deoxypalladtion releases the allylic benzoate **87aD**.

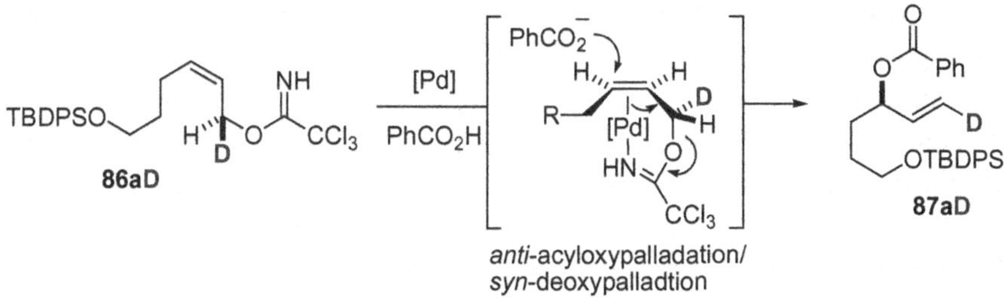

Scheme 4.97 Mechanism—antarafacial S_N2' displacement.

4.3.17 Reaction #27: Ruthenium-Catalyzed Redox Isomerization of Propargylic Alcohols

Professor Trost and Robert Livingston at Stanford University developed a method, by which a variety of propargylic alcohols (**92a**, **92b**) could be transformed to α,β-unsaturated aldehydes (**93a**, **93b**) in high yields (Scheme 4.98).[64] One of the salient features of the reaction was the stereospecificity: only the *trans*-isomer formed.

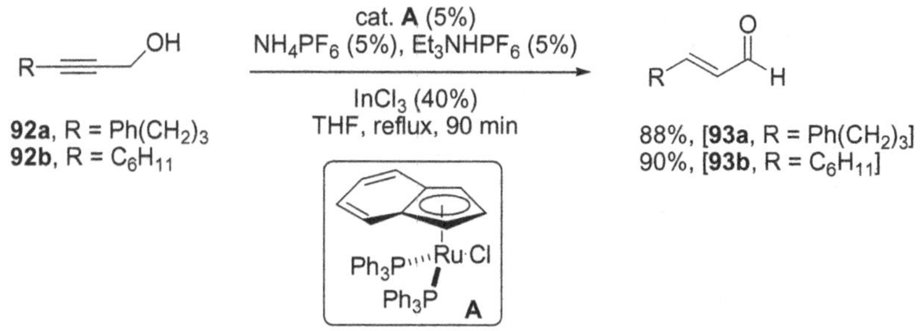

Scheme 4.98

Using the isomerization method, a highly efficient synthesis of leukotriene B_4 was achieved (Scheme 4.99). The key step of the synthesis was the isomerization of the propargylic alcohol **94** to the dienal **95**: under standard reaction conditions employing camphorsulfonic acid (CSA), the expected product **95** was obtained in 92% yield in 20 minutes of reaction time. Further reaction sequences afforded the target molecule.

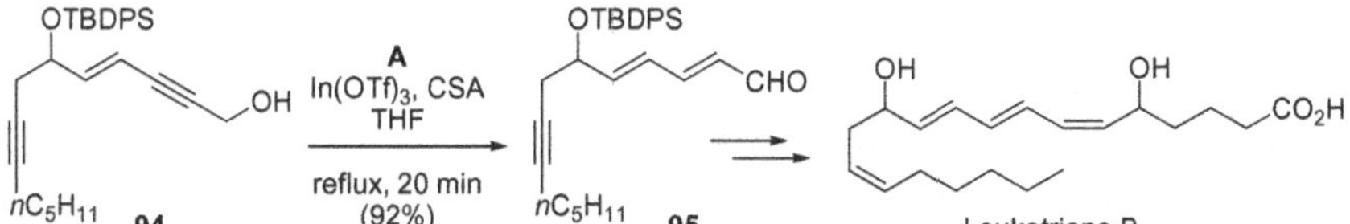

Scheme 4.99 Synthesis of leukotriene B_4.

To probe the reaction mechanism, deuterium-labeling experiments were performed (Scheme 4.100). The deuterium-labeled propargylic alcohol **96D2** was prepared by lithium aluminum deuteride reduction of methyl 2-tridecynoate. When the alcohol **96D2** was subjected to the standard reaction conditions, the expected aldehyde product **97D2** was obtained in 80% yield with complete deuterium incorporation both at the aldehyde and α-carbon. The result suggests that the product is formed by a net *anti*-hydrometalation. Another experiment using ordinary alcohol **96** and heavy water resulted in the formation of aldehyde **97D1**, which had 50% deuterium incorporation at the β-carbon.

$CH_3(CH_2)_9$—≡—CO_2Me → ($LiAlD_4$, THF, −78 to −30 °C; (44%)) → **96D2**

96D2 → (cat. **A** (5%), NH_4PF_6 (5%), Et_3NHPF_6 (5%); $InCl_3$ (40%), THF, reflux, 2 h) → **97D2** (80% y)

96 → (cat. **A** (5%), D_2O (1 eq); $InCl_3$ (40%), THF, reflux, 3 h) → **97D1** (33% y) (50% D)

Scheme 4.100 Deuterium-labeling experiments.

Another interesting experiment was a cross-over experiment where equimolar quantities of **92a** and **96D2** were subjected to the standard reaction conditions: both products **93a** and **97D2** were formed, and no deuterium incorporation was observed in the product **93a** (Scheme 4.101). This indicates that the hydride migration is indeed intramolecular, not intermolecular.

92a + **96D2** → (cat. **A** (5%), NH_4PF_6 (5%), Et_3NHPF_6 (5%); $InCl_3$ (40%), THF, reflux, 2 h) → **93a** + **97D2**

Scheme 4.101 Cross-over experiments.

Based on these experiments, the following mechanism was proposed (Scheme 4.102). Loss of chloride anion triggers the formation of reactive cationic ruthenium species ($\mathbf{A}^+$), which coordinates to the propargylic alcohol to produce the complex **B**. Intramolecular hydride transfer with concomitant loss of proton would give the vinylruthenium intermediate **C**. Reunion of triphenylphosphine with the intermediate **C** then delivers the product with a release of the active catalyst $\mathbf{A}^+$.

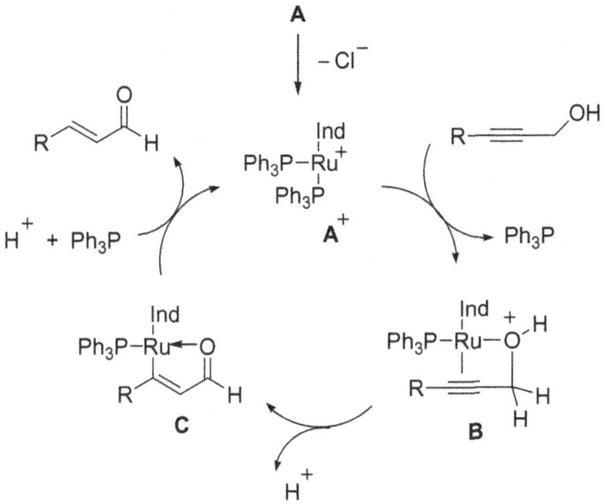

Scheme 4.102 Mechanism. Source: Reprinted from: Trost BM, Livingston RC. *J Am Chem Soc* 2008;**130**:11970. Copyright 2008 American Chemical Society.

4.3.18 Reaction #28: Aldol Reactions of Methyl Ketone Lithium Enolates

The aldol reaction of the lithium enolate of 3-methyl-2-butanone with chiral aldehyde **98** provided the product **99** of excellent diastereoselectivity (Scheme 4.103).

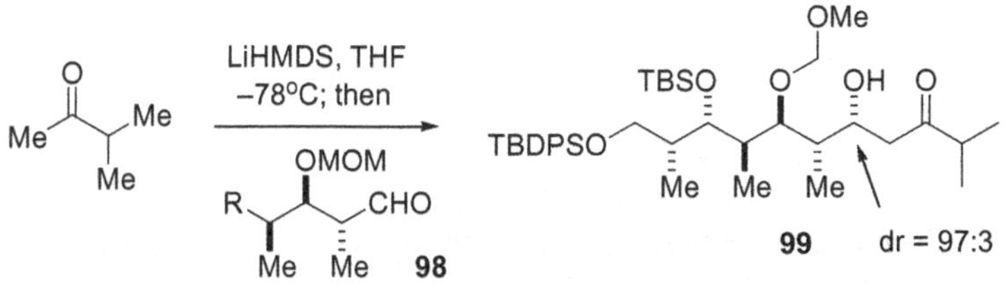

Scheme 4.103

The diastereoselectivity of the reaction had been rationalized by invoking a six-membered chair-like transition state. To further distinguish between the two competing chair-like and boat-like transition states, Professor Roush and coworkers at the University of Michigan carried out a mechanistic study employing deuterium-labeled enolsilanes (Scheme 4.104).[65] Deuterioboration of norbornadiene gave the exo-alcohol **100D**. Swern oxidation, addition of *tert*-butyllithium, followed by protection afforded the TMS ether **101D**. Finally, flash vacuum hydrolysis provided an inseparable mixture of deuterated enolsilanes (**102*E*** and **102*Z***) and a diprotio enolsilane (**102**H) in a 89:6:5 ratio.

NaBD$_4$, BF$_3$•OEt$_2$ THF, –78 to 23°C; NaBO$_3$•H$_2$O (68%)

100D

1. (COCl)$_2$, DMSO, Et$_3$N (87%)
2. *t*BuLi, –78°C (62%)
3. TMSCl, pyridine (72%)

101D

380°C, 2 mmHg (79%)

102*E* **102*Z*** **102**H (89:6:5)

*Scheme 4.104 Deuterium-labeled enolsilanes **102E** and **102Z**.*

With deuterated enolsilanes in hand, aldol reactions were carried out with benzaldehyde and chiral aldehyde **98** (Scheme 4.105). The isomeric purity of the enolsilanes (**102*E***, **102*Z***, **102**H) correlates very well with the ratio and pattern of deuterium labeling in the aldol products.

102*E* **102*Z*** **102**H (89:6:5)

MeLi, THF, 0–23°C; PhCHO –78°C (88:6:6)

MeLi, THF, 0–23°C; R'CHO (**98**) –78°C (91:5:4)

*Scheme 4.105 Aldol reactions of deuterium-labeled enolsilanes **102E** and **102Z**.*

The 2,3-*anti*-stereochemistry in the major aldol products is fully consistent with the dominance of a chair-like transitions state over a boat-like transition state (Scheme 4.106).[66]

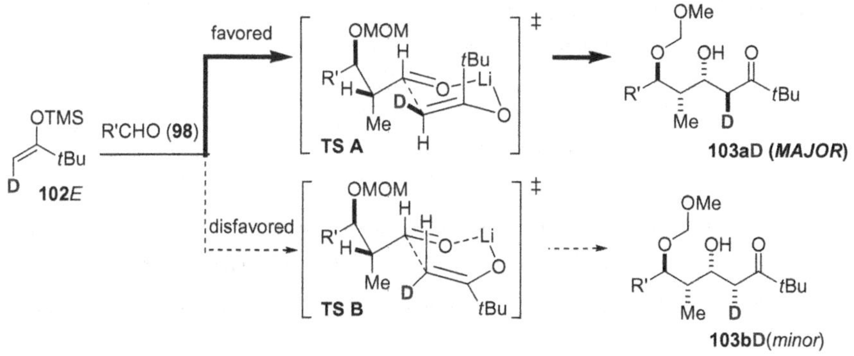

Scheme 4.106 Chair-like versus boat-like transition state.

The powerful use of deuterium was clearly demonstrated by the study. Without deuterium, it would have been impossible to tell the difference between two isotopic diastereomers **103aD** and **103bD**.

REFERENCES

1. Lewis GN. *J Am Chem Soc* 1933;**55**:1297.
2. Urey HC, Price D. *J Chem Phys* 1934;**2**:300.
3. Binder DA, Ellason R. *J Chem Ed* 1986;**63**:536.
4. Hughes ED, Ingold CK, Wilson CL. *J Chem Soc* 1934;493.
5. Knaus G, Meyers AI. *J Org Chem* 1974;**39**:1192.
6. Westheimer FH. *Chem Rev* 1961;**61**:265.
7. Wiberg KB. *Chem Rev* 1955;**55**:713.
8. Carey FA, Sundberg RJ. Advanced organic chemistry. *Part A: Structure and mechanisms.* Third Ed. New York: Plenum Press; 1990p. 216–20.
9. Urey HC. *Ind Eng Chem* 1934;**26**:803.
10. Hamasaki T, Aoyama Y, Kawasaki J, Kakiuchi F, Kochi T. *J Am Chem Soc* 2015;**137**:16163.
11. Robinson RK, De Jesus K. *J Chem Educ* 1996;**73**:264.
12. Reitz O, Kopp J. *Z Physik Chem* 1939;**A184**:429.
13. Streitwieser Jr. A, Heathcock CH. *Organic chemistry.* 2nd ed. New York: Macmillan Publishing Co., Inc.; 1981p. 376.
14. Hurd CD, Meinert RN. *Org Synth Coll* 1943;**Vol. 2**:541;
    Eisenbraun EJ. *Org Synth Coll* 1973;**Vol. 4**:310.
15. Westheimer FH, Nicolaides N. *J Am Chem Soc* 1949;**71**:25.
16. Fry A. *Chem Soc Rev* 1972;**1**:163.
17. Shiner Jr. VJ. *J Am Chem Soc* 1952;**74**:5285.

18. Saunders Jr. WH, Edison DH. *J Am Chem Soc* 1960;**82**:138.

19. Brown HC, Krishnamurthy S. *J Am Chem Soc* 1973;**95**:1669.

20. Brown HC, Zweifel G. *J Am Chem Soc* 1959;**81**:247;
Brown HC, Murray KJ. *J Am Chem Soc* 1959;**81**:4108.

21. Brown HC, Murray KJ. *J Org Chem* 1961;**26**:631.

22. NobelPrize.org. *Herbert C. Brown—biographical.*

23. Kabalka GW, Newton Jr. RJ, Jacobus J. *J Org Chem* 1978;**43**:1567.

24. Cope AC, Berchtold GA, Peterson PE, Sharman SH. *J Am Chem Soc* 1960;**82**:6370.

25. Roberts JD, Sheehan JC. *A biographical memoir*. Washington, DC: National Academy of Sciences; 1991.

26. Wolff ME. *Chem Rev* 1963;**63**:55.

27. Löffler K, Kober S. *Ber.* 1909;**42**:3431.

28. Titouani SL, Lavergne J-P, Viallefont PH, Jacquier R. *Tetrahedron* 1980;**36**:2961.

29. Corey EJ, Hertler WR. *J Am Chem Soc* 1960;**82**:1657.

30. NobelPrize.org. *Elias James Corey—biographical.*

31. Fieser LF, Novello FC. *J Am Chem Soc* 1942;**64**:802.

32. Houk KN, Lin Y-T, Brown FK. *J Am Chem Soc* 1986;**108**:554.

33. Evans DA, Hoveyda AH. *J Am Chem Soc* 1990;**112**:6447.

34. Yang J. *Six-membered transition states in organic synthesis.* Hoboken, NJ: Wiley-Interscience; 2008 [chapter 4].

35. Inoue S-I, Takaya H, Tani K, Otsuka S, Sato T, Noyori R. *J Am Chem Soc* 1990;**112**:4897.

36. NobelPrize.org. *Ryoji Noyori—bibliographical.*

37. Liang T, Zhang W, Chen T-Y, Nguyen KD, Krische MJ. *J Am Chem Soc* 2015;**137**:13066.

38. Yang J. *Six-membered transition states in organic synthesis.* Hoboken, NJ: Wiley-Interscience; 2008 [chapter 3].

39. Obligacion JV, Neely JM, Yazdani AN, Pappas I, Chirik PJ. *J Am Chem Soc* 2015;**137**:5855.

40. Curto JM, Kozlowski MC. *J Am Chem Soc* 2015;**137**:18.

41. Willoughby PH, Niu D, Wang T, Haj MK, Cramer CJ, Hoye TR. *J Am Chem Soc* 2014;**136**:13657.

42. Premuzic E, Reeves LW. *Can J Chem* 1962;**40**:1870.

43. Wu X, Fors BP, Buchwald SL. *Angew Chem Int Ed* 2011;**50**:9943.

44. Milner PJ, Kinzel T, Zhang Y, Buchwald SL. *J Am Chem Soc* 2014;**136**:15757.

45. Jensen KL, Standley EA, Jamison TF. *J Am Chem Soc* 2014;**136**:11145.

46. Mei T-S, Werner E, Burckle A, Sigman MS. *J Am Chem Soc* 2013;**135**:6830.

47. Hilton MJ, Xu L-P, Norrby P-O, Wu Y-D, Wiest O, Sigman MS. *J Org Chem* 2014;**79**:11841.

48. Brown HC, Brown CA. *J Am Chem Soc* 1963;**85**:1005.

49. Bromination De La Mare PBD, Dunn TM, Harvey JT. *J Chem Soc* 1957;923;
Iodination Grovenstein Jr. E, Kilby DC. *J Am Chem Soc* 1957;**79**:2972.

50. Giles R, Kim I, Chao WE, Moore J, Jung KW. *J Chem Ed* 2014;**91**:1220. <http://pubs.acs.org/doi/full/10.1021/ed500093g>.

51. Sevov CS, Hartwig JF. *J Am Chem Soc* 2013;**135**:2116.

52. Hoover JM, Ryland BL, Stahl SS. *J Am Chem Soc* 2013;**135**:2357.

53. Simmons EM, Hartwig JF. *Angew Chem Int Ed* 2012;**51**:3066.

54. Quasdorf KW, Huters AD, Lodewyk MW, Tantillo DJ, Garg NK. *J Am Chem Soc* 2012;**134**:1396.

55. *Chem Eng News*; March 9, 2015. p. 46.

56. Miyashita M, Sasaki M, Hattori I, Sakai M, Tanino K. *Science* 2004;**305**:495.

57. Mori K, Sueoka S, Akiyama T. *J Am Chem Soc* 2011;**133**:2424.

58. He Z, Yudin AK. *J Am Chem Soc* 2011;**133**:13770.

59. Li J, Burke MD. *J Am Chem Soc* 2011;**133**:13774.

60. Vougioukalakis GC, Grubbs RH. *Chem Rev* 2010;**110**:1746.

61. Stewart IC, Keitz BK, Kuhn KM, Thomas RM, Grubbs RH. *J Am Chem Soc* 2010;**132**:8534.

62. NobelPrize.org. *Robert H. Grubbs—biographical.*

63. Cannon JS, Kirsch SF, Overman LE, Sneddon HF. *J Am Chem Soc* 2010;**132**:15192.

64. Trost BM, Livingston RC. *J Am Chem Soc* 2008;**130**:11970.

65. Liu CM, Smith III WJ, Gustin DJ, Roush WR. *J Am Chem Soc* 2005;**127**:5770.

66. Yang J. *Six-membered transition states in organic synthesis.* Hoboken, NJ: Wiley-Interscience; 2008 [chapter 2].

CHAPTER 5

Applications in Medicinal Chemistry

5.1 BIOLOGICAL EFFECTS OF D_2O

Urey was very curious about heavy water[1]:

> *The biological interest of heavy water can hardly be overemphasized, since all living things live essentially in a water solution. It is my own expectation that both animals and plants can be acclimatized to high concentrations of heavy water, but that probably their living processes will be much slower.*

Many studies have been conducted with heavy water to test the biological effects on the plants and animals. It was Professor Gilbert N. Lewis at the University of California at Berkeley who conducted in 1933 the very first experiment on growing plants with D_2O. He wrote[2]:

> *I predicted that D_2O would not support life and would be lethal to higher organisms. As soon as heavy water became available, I began experiments to test this idea on the minute seeds of tobacco (*Nicotiana tabacum *var.* purpurea*). I placed twelve of these seeds in pairs in six glass tubes and added to each of three tubes 0.02 mL of ordinary distilled water and to each of the other three tubes 0.02 mL of heavy water. The six tubes were then hermetically sealed and placed in a thermostat at 25°C. The three pairs of seeds in ordinary water began to sprout in two days and at the end of two weeks formed the well-developed seedlings. The three pairs of seeds in heavy water showed no development.*

*Reprinted from: Lewis GN. J Am Chem Soc 1933;**55**:3503.*
Copyright 1933 American Chemical Society.

Deuterium. DOI: http://dx.doi.org/10.1016/B978-0-12-811040-9.00005-9

The marked difference between ordinary water and heavy water is striking, indeed.

A similar damaging effect of heavy water was observed on the growth of *Atropa belladonna* (Fig. 5.1).[3]

Increasing the deuterium concentration inhibited the growth: At concentrations up to 50%, flowering and berry formation occurred, but in the more highly deuterated plants, the berries were smaller and contained fewer seeds. At a concentration of 60%, no flowering occurred!

Although substantial replacement of hydrogen by deuterium was incompatible with higher plants, some organisms can grow in significant amount of D_2O. For example, the green algae *Chlorella vulgaris* and *Scenedesmus obliquus* were grown in media containing 99.7% D_2O with carbon dioxide as the sole carbon source. Bacteria and yeasts can be adapted to grow in heavy water.[3]

Some clinical studies were conducted to test the biological effects of D_2O on animals.[4] For example, Vasdev and coworkers found that inclusion of 10% D_2O in the drinking water of ethanol-fed rats lowered

Figure 5.1 Effect of D_2O concentration on plant growth. Reprinted with permission from AAAS: Katz JJ, Crespi HL. *Science* 1966;**151**:1187.

blood pressure.[5] Wallace and coworkers discovered that high D_2O concentrations might be valuable in boron capture therapy to treat brain tumors. In this treatment, the patient's brain was loaded with 40% heavy water during exposure to a neutron beam.[6]

5.1.1 Can We Drink Heavy Water?

Although heavy water has a taste not too different from ordinary water, it is not recommended to drink too much of it. Here is why. An extensive series of studies on the physiological effects of deuterium oxide in mice revealed that the acute lethal dose for heavy water was 5–7 mL per 10 g of body weight. However, a mouse ingesting 40% D_2O in its drinking water for a 2-month period showed no apparent ill effects other than retarded weight gain.

One interesting use of heavy water is in the determination of total body water in the human subject. By the method of deuterium dilution, Hevesy and Hofer in 1934 found a value of 63% in man.[7] A later study conducted at the Brigham Young Hospital in Boston found similar numbers: 61.8% for males and 51.9% for females.[8]

Subjects were normal medical students, college undergraduates, and hospital personnel in good health, ranging in age from 18 to 32 years. A total of 17 males and 11 females participated in the study. Heavy water (90 mL, 99.8% D) was filtered to an Erlenmeyer flask containing 0.7 g of NaCl. The solution was quantitatively mixed with normal saline and infused intravenously into the subject. After infusion, blood samples for total body water determinations were taken at least every 30 minutes for several hours. The whole blood was allowed to clot, and the serum was separated, analyzed. The serum was vacuum distilled twice, and the condensates were collected. Deuterium measurements both by falling drop technique and mass spectrometer were used to calculate the total body water.

5.2 DEUTERIUM IN MEDICINAL CHEMISTRY

In his acceptance speech of the twenty-third presentation of the Willard Gibbs Medal in 1934, Urey expressed his optimism about the beneficial uses of deuterium in medicine[1]:

The medicinal effects have often been mentioned, but mostly without adequate foundation. Experiments made on the effects of heavy water on cancer seem to indicate that there is little difference in the behavior of such tissue in the presence of either ordinary or heavy water. It may be, however, that medicinals, which will have valuable properties, can be prepared using deuterium.

Because the drug molecule is an organic compound that contains a number of C–H bonds, there is a chance that a deuterium kinetic isotope effect would be observed in the metabolism of drug. That is, if one of the C–H bonds of the drug molecule is broken in the metabolism, substitution of deuterium for hydrogen would slow down the metabolism of the drug by enzymes resulting in a longer action. Three early examples are presented here to demonstrate the effect.[9]

5.2.1 Metabolism Studies

In 1961, Rapoport and coworkers at the University of California, Berkeley, carried out a study to understand the metabolism of morphine. For the study, deuterium-labeled morphine **1D** was chosen (Scheme 5.1).[10] To prepare deuterated morphine **1D**, normorphine was treated with excess ethyl chloroformate in the presence of potassium hydroxide to give *O*^{3},*N*-dicarboethoxynormorphine. Lithium aluminum deuteride reduction of the carbamate afforded the deuterated morphine **1D** in 73% yield (Scheme 5.1).

Scheme 5.1 Deuterium-labeled morphine.

The pharmacological activity of both morphine and deuterated morphine determined on the swiss albino mice showed that deuterated morphine is less potent than the parent compound. For example, LD_{50} (**1D**) was 400 mg/kg and that of protio morphine was 256 mg/kg when both were injected subcutaneously, leading to a potency ratio of 1.56. The effect of deuteration was further tested in vitro by measurement of formaldehyde evolved from a fortified incubate of rat liver microsomes and morphine or deuteriomorphine (Scheme 5.2). Deuterated morphine was slower than the protio morphine, resulting in a kinetic isotope effect of $k_H/k_D = 1.4$. This value is comparable to the potency ratio of 1.56. The data prove that the rate of demethylation is dependent on the ease with which the C–H bond is oxidized and the biological actions are a function of such *N*-demethylation.

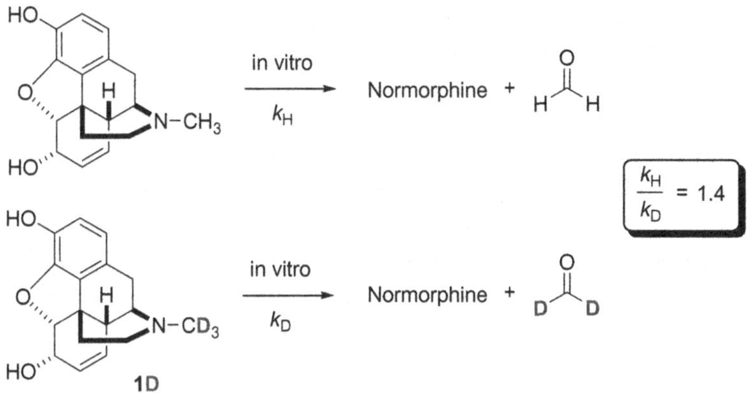

Normorphine + H(C=O)H, in vitro, k_H
1D: Normorphine + D(C=O)D, in vitro, k_D
$k_H / k_D = 1.4$

Scheme 5.2 Kinetic isotope effect.

In 1969, Tanabe and coworkers at the Stanford Research Institute wanted to test the pharmacological effect of deuterium substitution for hydrogen on drugs.[11] For the study, a sedative medicine butethal was chosen. Deuterium-labeled butethal **2D** was then synthesized starting from ethyl bromide-d_2 (Scheme 5.3).

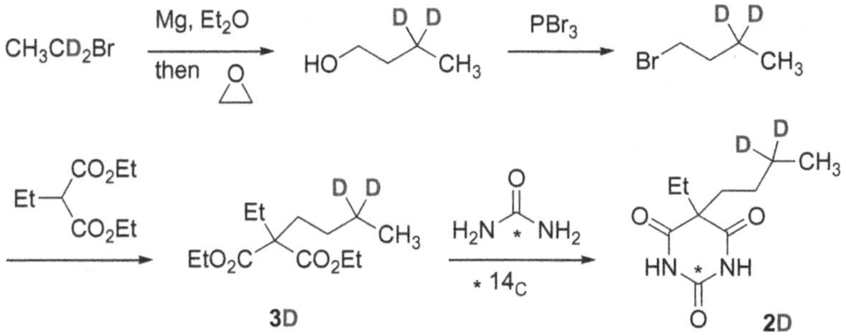

CH_3CD_2Br → (Mg, Et_2O, then ethylene oxide) → $HOCH_2CH_2CD_2CH_3$ → (PBr_3) → $BrCH_2CH_2CD_2CH_3$ → (Et-CH(CO_2Et)$_2$) → **3D** → (H_2N-CO-NH_2, * ^{14}C) → **2D**

Scheme 5.3 Deuterium-labeled butethal.

Reaction of Grignard reagent prepared from ethyl bromide-1,1-d_2 with ethylene oxide gave 1-butanol-3,3-d_2, which was converted to the corresponding bromide by the action of PBr_3. The C-alkylation of ethyl diethylmalonate with the bromide yielded the disubstituted malonate **3D**. Finally, condensation of the malonate **3D** with radiolabeled urea-C^{14} afforded 5-ethyl-5-(3'-dideuterio-*n*-butyl)barbituric acid-2-C^{14} **2D**. The use of radiolabeling was for the purpose of biological assays.

A change in the drug's activity was observed from an *in vivo* study with mice. More than a twofold increase in the sleeping time was induced by deuteriobutethal **2D** compared to the protiobutethal. As expected, the biological half-life of deuteriobutethal **2D** was approximately two and a half times longer than that of the protiobutethal. The metabolite of butethal was identified as 5-ethyl-5-(3'hydroxybutyl) barbituric acid (Scheme 5.4). The in vitro hydroxylation of protio- and deuteriobutethal revealed a kinetic isotope effect of 1.59, which suggests that the slower cleavage of C–D bond contributes to the longer half-life that results in a prolonged sleeping time of mice.

in vitro k_H

$\frac{k_H}{k_D} = 1.59$

in vitro k_D

Scheme 5.4 Kinetic isotope effect.

Sevoflurane is a halogenated general inhalation anesthetic drug. It is metabolized by cytochrome P450 2E1 enzyme to hexafluoroisopropanol, carbon dioxide, and fluoride anion (Scheme 5.5).[12]

P450

Sevofluorane

Scheme 5.5

In an attempt to increase the efficacy of the drug by slowing down the metabolism, Baker and coworkers invented a deuterium analog of sevofluorane (Scheme 5.6).[13] Treatment of hexafluoroisopropanol with dimethylsulfate-d_6 in the presence of base gave the methyl ether-d_3 in good yield. Fluorination using bromine trifluoride afforded sevofluorane-d_2.

$(F_3C)_2CH{-}OH + (CD_3)_2SO_4 \xrightarrow[\text{rt, 2h (77\%)}]{\text{10\% aq. NaOH}} (F_3C)_2CH{-}OCD_3 \xrightarrow[\text{(39\%)}]{BrF_3}$ Sevofluorane-d_2

Scheme 5.6 Sevofluorane-d_2.

The metabolism study was then conducted on both ordinary and deuterated sevofluorane using hepatic microsomes from male rats (Scheme 5.7). Measurement of the amount of the fluoride anion showed that defluorination was 17 times slower for deuterated sevofluorane than for ordinary sevofluorane, confirming that the C–H bond cleavage is involved in the rate determining step of the metabolism.

Sevofluorane $\xrightarrow[k_H]{\text{P450 in vitro}}$ F^- (1.36 nmol/mg protein)

Sevofluorane-d_2 $\xrightarrow[k_D]{\text{P450 in vitro}}$ F^- (0.08 nmol/mg protein)

$$\frac{k_H}{k_D} = 17$$

Scheme 5.7 Kinetic isotope effect in metabolism.

5.2.2 Recent Progress in the Development of Deuterated Drugs

Deuterium-labeled drugs have proved useful in understanding the mechanism of drug action and in elucidating metabolic and biosynthetic pathways.[14] Although deuterated drug molecules are useful as a biomarker, they have not been recognized as a new drug molecule different from their protio drugs.[15]

The situation may change, as a number of pharmaceutical companies have recently shown their interest in developing deuterium-labeled or deuterated drugs.[16] The main goal here is on the improvement of the drug's efficacy by substituting deuterium for protium in the already approved drugs.[17] This deuterium switching strategy has proven promising. Two examples will be presented here to highlight the fascinating applications of deuterium in medicinal chemistry.

5.2.2.1 Example 1: DeuteRx

CC-122, which has been developed by Celgene Corporation, is a pleiotropic pathway modulator that exhibits a broad range of activity such as teratogenicity and in vitro anti-inflammatory activity (Scheme 5.8). Although the (*S*)-enantiomer is presumed to be responsible for the medicinal activity, only a limited number of experimental data are available to support the conclusion due to the fact that the C-3 hydrogen is prone to in vivo racemization.

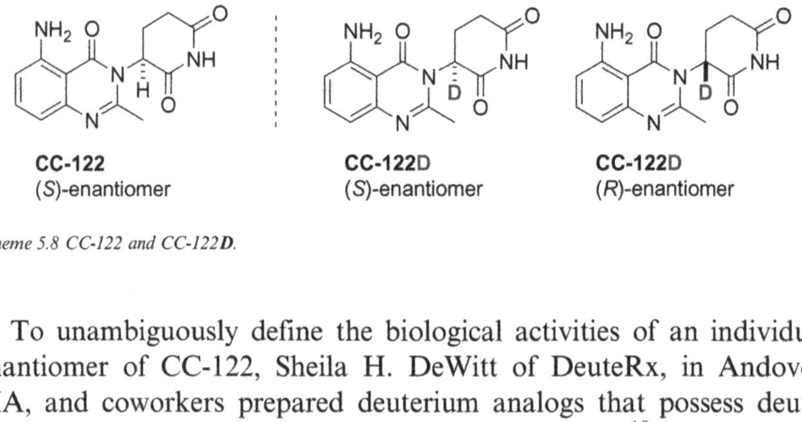

CC-122 (*S*)-enantiomer | **CC-122D** (*S*)-enantiomer | **CC-122D** (*R*)-enantiomer

*Scheme 5.8 CC-122 and CC-122**D**.*

To unambiguously define the biological activities of an individual enantiomer of CC-122, Sheila H. DeWitt of DeuteRx, in Andover, MA, and coworkers prepared deuterium analogs that possess deuterium at the C-3 position of the glutarimide moiety.[18] The rationale behind the strategy called deuterium-enabled chiral switching (DECS) was to exploit the nature of the C–D bond, which is stronger than the C–H bond.

Each enantiomer of the CC-122D was prepared starting from a coupling of the benzoic acid **4** with 3-aminoglutarimide (Scheme 5.9).

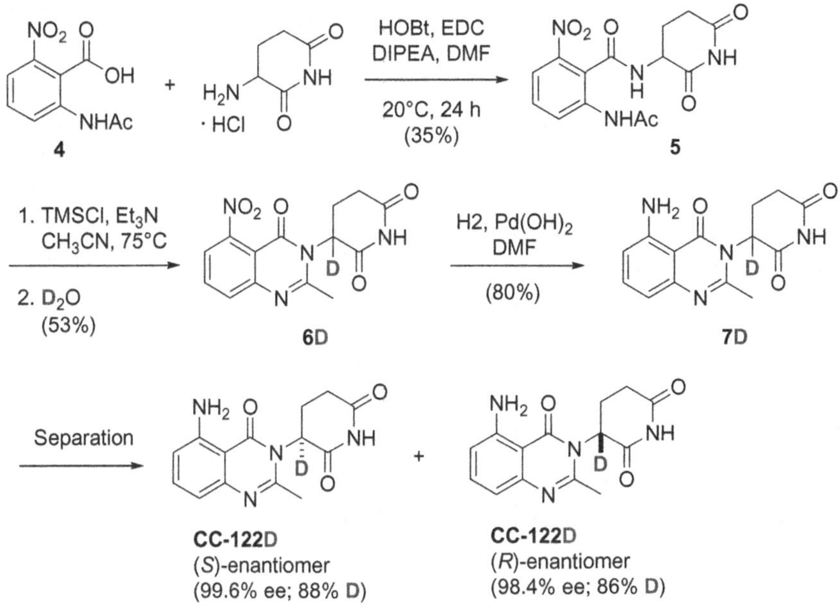

*Scheme 5.9 Synthesis of (S) and (R)-CC-122***D**.

The benzoic acid **4** was coupled with racemic 3-aminoglutarimide using [*N*-ethyl-*N*'-(3-dimethylaminopropyl)carbodiimide] and 1-hydroxybenzotriazole to give the amide **5**, which was converted to the mono-deuterated nitroquinazoline **6D** after a successive treatment with trimethylsilyl chloride/triethylamine and heavy water. Reduction followed by chromatographic separation afforded (*S*)- and (*R*)-enantiomers in pure form.

With both enantiomers in hand, a biological study was conducted to see if there was any difference between the two isomers in anti-inflammatory activity by measuring the ability to inhibit TNF-α production (Table 5.1). The test results show that the levorotatory isomer is 20-fold more potent at inhibiting TNF-α production than the dextrorotatory isomer.

Table 5.1 Anti-inflammatory Activities

Compound	IC_{50} (nM)
(−)-CC-122D	48.5
(+)-CC-122D	945

Chirality plays an important role in drug discovery. As of 2006, half of the new drugs contain at least one chiral center. Because stereochemical isomers exhibit different biological activities, pharmaceutical companies would like to develop a single enantiomer drug when possible.[19] When the chiral center of the drug molecule possesses a labile C–H bond subject to epimerization, it is not worth making an effort to synthesize a single enantiomer. The situation, however, suddenly changes with introduction of deuterium. It is now possible to stabilize a chiral center against epimerization.

The DECS strategy employed by researchers at DeuteRx opens up a new area, where developing a chiral drug is now possible by simply replacing a labile C–H bond with a stronger C–D bond.

5.2.2.2 Example 2: Concert Pharmaceuticals

The widely used antidepressant paroxetine also reduces hot flashes, but patients taking certain other drugs cannot use it because the P450 enzyme CYP2D6 metabolizes the methylenedioxy portion of the paroxetine to a carbene that then irreversibly inhibits the enzyme. As this liver enzyme is responsible for metabolizing up to a quarter of all medications, inactivating it with paroxetine can allow other medications a patient is taking to build up to dangerous levels in the blood stream.

Concert Pharmaceuticals of Lexington, MA, developed CTP-347, a deuterium version of paroxetine, in an effort to reduce the formation of this carbene metabolite (Scheme 5.10).[20]

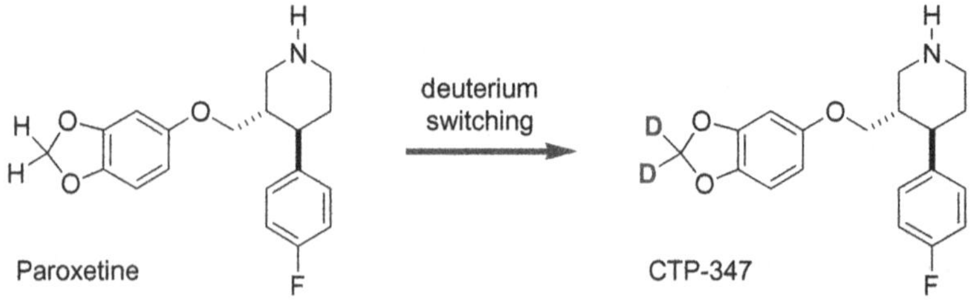

Scheme 5.10 Paroxetine-d_2.

Reaction of 3,4-dihydroxybenzaldehyde-d_2 **8** with CD_2Cl_2 gave piperonal-d_2 **9D** in good yield (Scheme 5.11). A sequence of Bayer–Villiger oxidation and hydrolysis afforded sesamol-d_2 **10D**.

Reaction of sesamol-d_2 **10D** with the mesylate **11** followed by demethylation gave the carbamate **12D** in a reasonable yield. Finally, saponification and hydrochloride salt formation delivered the target molecule, CTP-347.[21]

Scheme 5.11 Synthesis of CTP-347.

The pharmacological activities of CTP-347 were assessed compared to paroxetine using an in vitro rat synaptosome model. The IC_{50} levels with CTP-347 for serotonin and dopamine uptake were similar to those of paroxetine, confirming that deuterium substitution has little impact on the pharmacological activity (Table 5.2).

When tested for metabolic stability against human liver microsomes, CTP-347 was cleared faster than paroxetine. The higher rate of CTP-347 metabolism was due to a decrease in metabolic intermediate

Table 5.2 Reuptake Inhibition (IC_{50}, nM)

Neurotransmitter	CTP-347	Paroxetine
Serotonin	0.89	0.72
Dopamine	270	230

complex (MIC) formation with CYP2D6. This conclusion was further supported by a study conducted on in vitro metabolic effects of CTP-347 with tamoxifen. The study found that the administration of CTP-347 at a various range of concentration caused little or no change in the metabolism of tamoxifen, whereas the metabolism of tamoxifen decreased with high paroxetine concentrations. Thus CTP-347 effectively reduces drug–drug interactions with tamoxifen by helping to preserve CYP2D6 function.

The study has clearly demonstrated a powerful impact of deuterium: a simple substitution of deuterium for hydrogen can improve the safety and efficacy of existing therapeutic agents.

REFERENCES

1. Urey HC. *Ind Eng Chem* 1934;**26**:803.
2. Lewis GN. *J Am Chem Soc* 1933;**55**:3503.
3. Katz JJ, Crespi HL. *Science* 1966;**151**:1187.
4. Kushner DJ, Baker A, Dunstall TG. *Can J Physiol Pharmacol* 1999;**77**:79.
5. Vasdev S, Gupta IP, Sampson CA, Longerich L, Patai S. *Can J Cardiol* 1993;**9**:802.
6. Wallace SA, Mathur JN, Allen BJ. *Med Phys* 1995;**22**:585.
7. Hevesy G, Hofer E. *Nature* 1934;**134**:879.
8. Schloerb PR, Friis-Hansen BJ, Edelman IS, Solomon AK, Moore FD. *J Clin Invest* 1950;**29**:1296.
9. Blake MI, Crespi HL, Katz JJ. *J Pharm Sci* 1975;**64**:367.
10. Elison C, Rapoport H, Laursen R, Elliott HW. *Science* 1961;**134**:1078; Elison C, Elliott HW, Look M, Rapoport H. *J Med Chem* 1963;**6**:237.
11. Tanabe M, Yasuda D, LeValley S, Mitoma C. *Life Sci* 1969;**8**:1123.
12. Williams DA, Lemke TL. *Foye's principles of medicinal chemistry*. Fifth ed. Philadelphia: Lippincott Williams & Wilkins; 2002 [chapter 14].
13. Baker MT, Ronnenberg Jr WC, Ruzicka JA, Chiang C-K, Tinker JH. *Drug Metab Dispos* 1993;**21**:1170.
14. Nelson SD, Trager WF. *Drug Metab Dispos* 2003;**31**:1481.
15. Gant TG. *J Med Chem* 2014;**57**:3595.
16. Timmins GS. *Expert Opin Ther Pat* 2014;**24**:1067.
17. Yarnell AY. *Chem Eng News* 2009;**87**:36; Sanderson K. *Nature* 2009;**458**:269.
18. Jacques V, Czarnik AW, Judge TM, Van der Ploeg LHT, DeWitt SH. *Proc Natl Acad Sci* 2015;**112**:E1471. Chem Eng News, 2015, March 16, p. 26.
19. Farina V, Reeves JT, Senanayake CH, Song JJ. *Chem Rev* 2006;**106**:2734.
20. Tung, R. US Patent 7678914 B2; 2010.
21. Uttamsingh V, Gallegos R, Liu JF, Harbeson SL, Bridson GW, Cheng C, Wells DS, Graham P, Zelle R, Tung R. *J Pharmacol Exp Ther* 2015;**354**:43.

CONCLUSION

In previous chapters, I recounted a history of deuterium. I also presented a number of applications in organic and medicinal chemistry research areas to showcase the mighty uses of deuterium.

A final note that I would like to make is about a practical use of deuterium in the area of analytical chemistry. In measuring the minute amount of a specific chemical in the environment or in our body, deuterium-labeled compounds are used as an internal standard. The method being used by researchers is called stable isotope dilution (SID).[1] In this method, a known concentration of deuterium-labeled compound is spiked into a solution of unknown concentration. By using gas chromatography or liquid chromatography tandem mass spectrometry (GC- or LC-MS/MS), an accurate determination of an analyte concentration can be made through an analysis of mass spectrum. A wide variety of deuterium-labeled compounds are available to researchers for the purpose of SID analysis. Three compounds are shown here to illustrate their uses (Scheme 1).

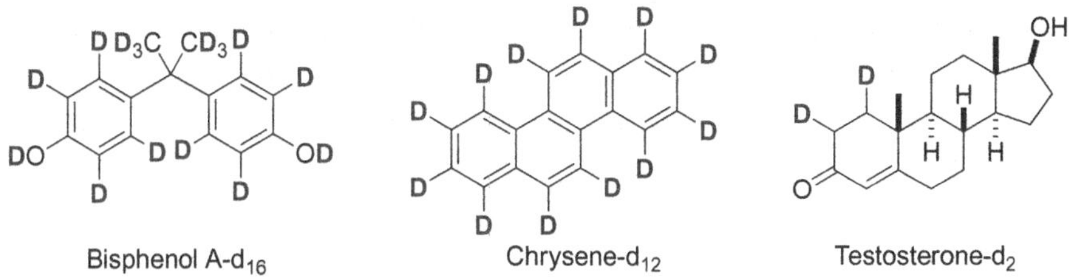

Bisphenol A-d_{16} Chrysene-d_{12} Testosterone-d_2

Scheme 1

Bisphenol A (BPA) is used in the manufacturing of epoxy resins and polycarbonate plastics, which are constituents of a variety of products including plastic food containers and water bottles, and dental fillings. As BPA can leach into food and beverages from plastic containers, humans are constantly exposed to BPA. The recent ban in 2012 by the Food and Drug Administration on the use of BPA for baby bottles highlights the public's concern about the toxicity of BPA.

Because BPA is an endocrine disrupting chemical, many studies have been conducted using bisphenol A-d_{16} as an internal standard to accurately determine the amount of BPA in human. For example, Woodruff and coworkers at the University of California, San Francisco, determined the amount of BPA in human umbilical cord serum.[2]

Chrysene, found in coal tar, is being used in the manufacture of coatings, dyestuffs, and road paving. This chemical belongs to a group of polyaromatic hydrocarbons (PAHs), which are classified as a probable human carcinogen by Environmental Protection Agency (EPA). Lohmann and coworkers at the University of Rhode Island used chrysene-d_{12} to determine the amount of chrysene in the Narragansett Bay area.[3]

Testosterone is a potent androgen that has important sexual and metabolic activities. Measurement of testosterone in serum or plasma is essential in the investigation of androgenic status and monitoring of replacement therapy in children and adult of both sexes. Cawood and coworkers at the Leeds General Infirmary, UK, employed testosterone-d_2 to accurately measure the amount of testosterone by using only 50 μL of serum sample.[4]

Urey predicted that deuterium would have a far-reaching influence on many areas of chemistry. He was absolutely correct. The applications of deuterium seem to be limitless and I am sure Urey would be more than happy to see all the advances in chemistry made possible solely because of deuterium. It is without a doubt that deuterium will continue to be a beloved isotope in the future.

REFERENCES

1. Rychlik M, Asam S. *Anal Bioanal Chem* 2008;**390**:617; Ciccimaro E, Blair IA. *Bioanalysis* 2010;**2**:311.
2. Gerona RR, Woodruff TJ, Dickenson CA, Pan J, Schwartz JM, Sen S, et al. *Environ Sci Technol* 2013;**47**:12477.
3. Lohmann R, Dapsis M, Morgan EJ, Dekany V, Luey P. *Environ Sci Technol* 2011;**45**:2655.
4. Cawood ML, Field HP, Ford CG, Gillingwater S, Kicman A, Cowan D, et al. *Clin Chem* 2005;**51**:1472.

AUTHOR INDEX

Note: Page numbers followed by number within parenthesis denote reference number.

Printed in the United States
By Bookmasters